A short history of Science

Frontispiece   Survey-scribe: statue of Senenmut,
architect of Queen Hatshepsut

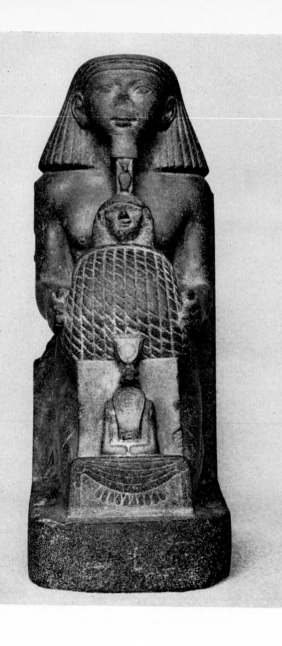

A short history
of Science

J. G. CROWTHER

Methuen Educational Ltd

LONDON · TORONTO · SYDNEY · WELLINGTON

First published in 1969
by Methuen Educational Ltd
11 New Fetter Lane, London E C 4
© 1969 J. G. Crowther
Printed in Great Britain by
Richard Clay (The Chaucer Press), Ltd
Bungay, Suffolk

Contents

Acknowledgements

The figures are reproduced by courtesy of the following:

2 La Musée du Louvre; 3 and 4 Dr Joseph Needham; 5, 6, 10, 14, 17, 19, 20, 22, 23, 28, 29, 33, 36 and 37 the Trustees of the Science Museum; 7, 9, 12, 13, 16, 24, 31 and 35 the Trustees of the British Museum; 8, 25 and 26 the Wellcome Trustees; 15 and 39 the Royal Institution; 34 Professor M. H. F. Wilkins; 38 the European Organization for Nuclear Research; 40 the Society for Cultural Relations with the USSR.

1. How science came into existence

The fossils of creatures that lived at least a million years ago, and may have been ancestors of man, have been found in various parts of the earth. These early sub-human creatures were already using stones as tools. Half a million years later their descendants, who lived in Java, China, Algeria and elsewhere, used flint implements and fire. Three hundred thousand years later still, that is about two hundred thousand years ago, there were even more advanced creatures, whose fossil skulls show that their brains were larger and more complicated. Their stone tools were more varied and better made.

Successors presently appeared, who began to bury their dead in definite ways, suggesting that they had ideas, for a particular mode of burial presumably had some kind of mental significance to it. Elaboration of burials began about fifty thousand years ago. This is exemplified by their arrangement in special grounds, with the accompaniment of carvings and paintings.

Until about ten thousand years ago men were preoccupied with hunting and war, and the particular kinds of implements and weapons required for them. Hunting and food-gathering were followed by settled forms of living, which involved the creation of a self-supporting and repetitive system of food production, based on the domestication of animals and the cultivation of plants.

Life in settlements stimulated a new order of invention. Men discovered how to make pottery out of clay, and they invented the potter's wheel. They discovered how to extract certain metals from ores, and work them into tools. The first historical records to throw light on these constructive activities begin only about five thousand years ago. They are associated with the growth of city life, which developed out of the earlier settlements and provided a still higher stimulus, leading to the invention of mathematics and writing. These in turn raised the fertility of invention to a new degree.

The historical age extending up to the present is witnessing an ever-increasing rate of invention and discovery, in which antibiotics, electronic computers, nuclear energy and space travel have appeared within the last thirty years. These spectacular advances, which seem at first sight to belong to a different order of existence, and to have little to do with the prehistoric past, are nevertheless rooted in what man did before recorded history. His ancestors' groping attempts to use stones as tools led them to go through those trials and motions which, hundreds of thousands of years later, became perfected in the operations of experimental science.

The effort to co-ordinate the actions of eye and hand, which is a kind of primitive scientific experimental activity, was one of the causes of the growth of the brain, which converted man's ancestors from animals into men. In a sense, science is older than man, and the attempt on the part of animals to be scientific was a cause of their evolution into men.

Prehistoric man recognized many of the basic facts on which modern science is based. He learned hundreds of thousands of years ago how to distinguish flint, which made the best implements, from other stones, and thereby acquired rudimentary notions of mineralogy. Towards the end of the Stone Age men dug mines fifty feet deep to obtain flint for tools. Knowledge of plants and animals was essential in order to obtain food. This was subsequently to form the basis of the sciences of botany and zoology.

There are more than two thousand species of edible plants. Prehistoric man and, perhaps even more, prehistoric woman, had to learn which they were, and which were worth gathering and storing, such as fruits, nuts and seeds. He had a considerable knowledge of plants when, about ten thousand years ago, he invented agriculture by cultivating some of them. Among these were wheat and rice. Altogether, prehistoric man cultivated more than two hundred species of plants.

His knowledge of animals is indicated by deposits of bones near his habitations, and by actual pictures. The cave pictures of animals, at least twenty thousand years old, include drawings of mammoths, reindeer, horses, cats, bears, boars, bison and rhinoceros. They show accurate biological observation, besides artistic merit.

Prehistoric man's medical knowledge is illustrated by his recognition of the heart, recorded in drawings of mammoths in which its shape and location are indicated. Prehistoric surgeons made tre-

I. Reproduction of a cave-drawing
of elephant and bison indicating the position of the heart

panning operations on the human skull, in which circular pieces of bone about the size of a penny were cut out with sharp flint knives. The patient could recover from the operation, for skulls have been found on which the operation had been performed several times in succession.

Prehistoric men had already made progress in arithmetic more than ten thousand years ago. They made rows of notches on stones or bones, in order to count the number of animals in herds, and other items that they possessed.

Prehistoric man lived much in the open, and observed nature intently. The sun and the stars were very much his companions, and already at a remote period he detected some connection between them and the seasons. The invention of agriculture stimulated his study of the stars, for this provided him with the beginning of a calendar to determine the best times for sowing and harvesting.

Science is older than history. The chief data on which it is based were contemplated by man and his ancestors for tens and hundreds of thousands of years before writing was invented. In fact, symbols for numbers were invented before writing. The first thing to recognize about science is that it is inherent in the oldest parts of human achievement, and in man himself. There would never have been any men, if their sub-human ancestors had not turned themselves into men by pursuing what, in essence, were primitive forms of scientific activity: the mastering of themselves and their environment.

As man pursued his struggle with himself and nature, his activity contained aspects which gradually developed into technology, pure science and applied science. In the beginning, all of these were merely aspects of one activity, of man's turning this way and that, trying to gain command of his surroundings in order to live and be happy. As he learned more about these aspects, it became convenient and necessary to regard each of them as a self-contained subject, to be distinguished from the others.

All the sciences sprang from the same root. When their common origin is obscured or forgotten man creates difficulties for himself.

2. The raw material of science

Ten thousand years ago the climate of large regions of the earth was different from what it is today. The northern lands were colder, and England had only recently become free from glaciers. Areas in North Africa were cooler and damper, and were able to support prehistoric agricultural settlements. As the climate grew warmer these areas became drier, and ultimately desert. The peoples formerly scattered over them were driven towards the valley of the Nile, which became the only part able to provide food. Similar movements occurred in Mesopotamia, India and China.

The result of this development was that, about seven thousand years ago, peoples who had already achieved a variety of skills in making tools and cultivating plants and animals found themselves contained in several of the great river valleys of the world. They had very fertile soil near the rivers, but beyond these narrow areas they were encircled by expanses of almost impassable desert. This acted as a strong defence against invaders. In their comparatively secure isolation, these peoples were thrown in upon themselves, and were able to pursue their lives without interruption. They developed their agricultural pursuits along the lines facilitated by the valley conditions. Authorities are not agreed in which valley these developments first occurred, but here that of the Nile is taken as the example.

In the Nile the annual floods left deposits of fertile silt, on which luxuriant crops could be grown. The Egyptians dug channels and built banks to divert the silt-bearing flood waters over further areas, to increase the production of grain. The channels and banks became more complex and bigger, and required more advanced skill and technique for their construction. It was from their experience in constructing these water-works that the ancient Egyptians discovered the basic principles of engineering; this enabled them to design and build their great pyramids, that provide such impressive proof of their genius.

The concentration and uniformity of conditions in the Nile valley were favourable to unification. Peoples who had some thousand years before lived in independent settlements scattered over prairies were now brought into continuous contact, and ultimately under unitary control.

The effective history of Egypt begins with the king Menes, who combined the upper and lower parts of the valley into one realm, about five thousand years ago. During the next thousand years, up to about 2000 B C, the great pyramids were built, and the foundations of Egyptian science and technology laid. In the first place, measuring and surveying of the land had to be invented in order to plan the agricultural system, based on the control of flood waters. The floods temporarily eliminated the landmarks, so that farmers no longer knew precisely where their farms were. Exact measurements of fields were required, so that the position of farms could be located without dispute after the waters had receded.

The measuring of the areas of land led to the invention of geometry. Later on this geometry, which originally arose out of irrigated agriculture, was utilized in the building of banks to control the flood water, and applied later in the construction of pyramids.

The ancient Egyptians developed arithmetic to measure the volume of their grain harvests and allocate it among the people. As the whole country and everything in it belonged to the king, there was no private property. The king and his priestly clerks had to calculate how much each man should have. Ancient Egyptian society had a hierarchical structure. The people were rigidly divided into classes, the members of each of which were entitled to their particular quantity of grain.

The arithmetic which the Egyptians elaborated was suited to making this kind of calculation. Multiplications and divisions were made with the aid of repeated doublings. They never multiplied by more than two in one operation. They recorded their results with the aid of a decimal system of signs. Their method of computation was very simple and made little demand on the memory, but, like the modern computer working on a scale of two, its operations were tedious.

The Egyptian method of calculating the area of a triangular piece of land led them to a good method of calculating the area of a circle. They did this by inscribing it in a square, and estimating the difference by drawing triangles whose areas could be calculated. They obtained the value of 3·1605 for π, the ratio of the circumference of a circle to its diameter.

The Egyptians advanced in other sciences besides mathematics and engineering. They had the most accurate of the ancient calendars, and their year contained 365 days. Their astronomy enabled them to set the face of the Great Pyramid to the north within an accuracy of one-twentieth of a degree. Indeed, their development of the sense of accuracy in measurement and construction prepared the way for the conception of *proof* in mathematics and science, which was elaborated by the Greeks.

The ancient Egyptians were distinguished in surgery. They used lint, and bandages of an adhesive character. They supported fractured limbs with wooden splints bound with bandages, and used other devices to hold wounded parts steadily in the best position. Their diagnosis and treatment were in keeping with such techniques. They practised extensive dentistry, and instances of a bridge to support a loose tooth, and the draining of an abscess under a molar tooth by drilling a hole in the jawbone, are known. Their medicines included castor oil and various materials which contain important therapeutic substances. They treated diseases of the eye with bile, from which cortisone can be derived. They used bats' blood and liver preparations, which are rich in vitamin A. Some or all of these may have been handed down from prehistoric magicians, whose brews had been shown by experience to have curative value.

The ancient Egyptians left descriptions of concrete examples of their mathematics and medicine carved on stone, and subsequently on sheets made from the papyrus plant, a reed growing in the river marshes.

The Mesopotamians, who were progressing in the valley of the Euphrates and Tigris, invented another form of written records more than five thousand years ago. The shortage of stone in the valley led to the use of clay for many purposes. They made wedge-shaped marks with a sharpened reed on pieces of soft clay, which were subsequently baked. Hundreds of thousands of such clay tablets, bearing records of this cuneiform script, have survived.

As with the Egyptians, Mesopotamian science grew out of the struggles with the problems of daily life. They surpassed the Egyptians as calculators, inventing more subtle techniques. Their numeral system was based first on ten, and then on six. Their division of the circle into six times sixty gradations is the origin of the angular degree which is still in use. No satisfactory explanation of why they adopted the scale of six has yet been found. They had only two number symbols, for one and ten. This made their arithmetical symbolism clumsy. But the Mesopotamians introduced the system of place-value, by which the same symbol may have different values according to its place. Thus the system still used today, by which the symbol for one may represent the number one, or ten, or a hundred etc, according to its place, has been inherited from the Mesopotamians.

The subtlety that they exhibited in arithmetic was reflected in their algebra. They succeeded in finding the solution of equations of the first, second and third degree.

They made less progress in geometry. They solved, or tried to solve, geometrical problems, such as the calculation of areas, by entirely arithmetical methods. They often accompanied such calculations with drawings to represent the area to be calculated, but this was diagramatic and not in proportion.

The preference of the Mesopotamians for arithmetic and calculation was influenced by the materials and conditions in the Euphrates valley. Unlike Egypt, it was lacking in stone, which is inherently so geometrical in its rigidity. Clay, which is inherently shapeless and ungeometrical until changed by baking, was the staple material for both buildings and the tablets for writing and calculation. The natural phenomena of the Euphrates valley were much less regular than in the valley of the Nile. The floods were less predictable. Unlike the Egyptians, who found permanence in stone and the regularity of the floods, the Mesopotamians were forced to look for it elsewhere, and they believed that they had found it in numbers and the regularities of arithmetic.

They were very industrious compilers of facts and figures. They made comprehensive tablets of weights and measures, and numbers such as squares and cubes. They calculated cube roots, and they stated the value of the square root of 2 correct to five places of decimals. They compiled arithmetical progressions, and stated what their sums would be to a particular number of terms. They left tables of a form of logarithm. They did not record general methods of solving algebraic equations, but the many and various correct solutions they discovered imply that they had some kind of understand-

2. Babylonian astrological tablet

ing of general method. It is possible that this part of the mathematical technique was handed down verbally. On the evidence of what has survived, algebra may be regarded as a product of the Mesopotamian tradition.

Their astronomy was very numerical in character. It was distinguished by predictions of eclipses, which they could calculate from the results of long-continued careful observations of the moon and sun. In this field they quite surpassed the Egyptians. They did not, however, visualize the machinery of the heavens in geometrical

B

terms. They were interested in exact observations, and then using their calculating powers to predict from them what would happen, but not how it would happen.

The ancient Egyptians and Mesopotamians, and in a comparable degree also the ancient Indians and Chinese, accumulated much exact observation of many aspects of nature. Altogether the first three thousand years of history, up to about 500 BC, had produced a large and varied amount of work which contained an inherent scientific element. It provided an adequate accumulation of knowledge for critical reflection.

The ancient Greeks made remarkable progress in extracting the general principles which were implicit in it. They thereby became the founders of science in a form that is easily recognized as such by people today. They were the effective inventors of generalized thinking, that is, of thinking that applies to all or many cases, and not just to one particular case. They were, however, indebted to the Egyptians and Mesopotamians for much of the data from which they drew their general ideas and principles.

3. Greece and the formulation of basic scientific ideas

The Greeks were originally barbarians who had come from South Russia into Asia Minor, or Ionia, 'the land of the wanderers'. There they found themselves on the trade routes of Egypt and Mesopotamia. They had emerged comparatively recently from Stone Age agricultural life on the steppes. Their social organization was simpler and less rigid than that of Egypt and Mesopotamia.

Some of the Greeks found their way to Thebes and Babylon, and viewed the prodigious works and accomplishments. They were profoundly impressed, but not overawed. They returned to the Greek settlements, and reflected on what they had seen. Herodotus, Hippocrates, Aristotle and other great Greeks acknowledged their debt to the ancient civilizations, and it is sufficient to accept their acknowledgement at its face value.

In particular, the Greek visitors to Egypt returned with a knowledge of the geometry which the Egyptians had invented in the course of building their great works. They did not look at this geometry in the same way as its inventors. The Egyptians supposed that the mummified kings continued to watch over their welfare, and it was therefore necessary that they should be preserved as permanently and splendidly as possible. They regarded the great pyramid as the guarantee of the people's future. Its construction was the most important task, and might justifiably absorb all resources surplus to those required for subsistence.

The Greeks also wished to construct works in their homeland, but without this obsessive concentration, derived ultimately from the narrow conditions in the valley of the Nile. They built, but kept their constructions in reasonable proportion to the other interests of life. They devoted themselves to reflection on the ancient techniques, as well as using them. They were less concerned with the quest for eternal life in imperishable constructions, and paused to consider what they were doing and how they did it.

Thales was the first of the Greek travellers to the ancient civilizations to leave examples of the new attitude to science, the study of the subject itself, separated from the constructive objects for which it had originally been invented.

He was a native of Miletus on the coast of Asia Minor, where he was born about 630 BC. He was a merchant engaged in trade in salt and oil, and a naturally ingenious man. One of his mules carrying a load of salt stumbled in a stream. The salt dissolved in the water, and the mule felt its load reduced. After this, it made a habit of lying down in all streams. Thales cured it by substituting loads of sponges for salt. He was equally clever at outwitting men. Aristotle says that when he heard that there would be a glut of olives he cornered all the olive-presses in the district. When the glut came, and the demand for presses became heavy, he rented them out at a high rate. Besides being a smart merchant, Thales was active in local affairs.

He probably visited Egypt for commercial reasons. While he was there he became acquainted with Egyptian geometry. After he had returned to Miletus he reflected on the geometrical facts that he had learned. The first of these was that the angles at the base of an isosceles triangle, that is, a triangle with two equal sides, are equal. The Egyptians knew this by induction, as a result of much experience in building. Thales, not being an Egyptian temple-builder, looked for a shorter method of proof. He made two equal isosceles triangles, and laid one exactly on the other. He showed that if the top one were picked up and turned over it still fitted exactly when laid again on the lower one. The two angles at the base were therefore equal.

This demonstration depends on deduction; it might have been made even if these isosceles triangles had been the first ones in the history of the world. It was not logically dependent on the experience of the Egyptians during thousands of years. Nevertheless, Thales invented, or left the earliest recorded example of, deductive science, through his acquaintance with Egyptian science. The deductive method is logically distinct from the inductive, but is indissolubly bound up with it historically and socially.

Thales recognized that when two straight lines cross each other the vertically opposite angles were equal. He may have proved this by turning over a pair of crossed sticks which had been tied together. He proved that if the base of a triangle and the angles at its ends are given, then the triangle can have only one shape: its form is determined. He used the proposition to measure the distance from land of a ship at sea. He proved that the sides of equiangular triangles of different size are proportional in length, and was reputed to have measured the height of an Egyptian pyramid by the aid of this proof, in the presence of the king Amasis. He stated that a circle is cut in half by any diameter, and that the lines joining any point on the circumference of a circle to the ends of any diameter are at right angles to each other.

Particular cases of the last proposition could be seen in tiled floors and other decorations consisting of circles and squares, but the generalization of it marks a great advance in abstract thinking.

Thales was also eminent as an astronomer. He foretold an eclipse, no doubt on the basis of Mesopotamian data. According to modern calculations, it was probably the one that occurred on either September 30th, 609 BC, or on May 28th, 585 BC. It is said that once, when he was staring intently at the stars while walking out at night, he tripped up and fell into a ditch; whereupon an old woman asked him how he could discover what was in the sky if he could not see things lying at his feet.

He is credited with the first conception of the universe as being made out of one primary element, which formed the objects of nature by passing through a series of transformations. He suggested that this primary element was water. He said that the earth is borne on water, and that the fire of the sun and stars was surrounded by exhalations of water. He was perhaps reflecting on what sort of mechanism would be necessary to explain the Mesopotamian stories of the origin of the world, one of the versions of which is incorporated in the Book of Genesis.

Thales was followed by two other notable Milesians, Anaximander and Anaximenes. They extended the notion of a primary element for explaining physical phenomena. Anaximander applied the idea of transformations of a primary material to explain the origin of the earth and stars, and living organisms as the product of transformations of the earth. He suggested that life began in water, under the effect of evaporation from the sun's heat, and emerged on to the land, to which it began to adapt itself. He said the ultimate ancestor of man resembled a fish.

Pythagoras was born at Samos, off the coast of Asia Minor, and was about thirty years younger than Thales. It is said that he became a pupil of Anaximander, who recommended him to study in Egypt. Whether or not this is true, Pythagoras showed himself as much a follower of the Mesopotamians as the Egyptians. He was profoundly impressed with numbers and calculation. He regarded numbers as real things, and the substance out of which material objects were made. The extraordinary importance that Pythagoras ascribed to numbers arose from deep causes. It is not without significance that it was in Asia Minor, in Pythagoras's time, that money first began to supersede barter as a means of trade and exchange; it stimulated the expression of values in numerical terms, and enhanced the status of numbers.

Pythagoras's teaching appealed to the upper classes. He formed a secret brotherhood, whose members lived a simple life, and engaged in research and meditation. His organization acquired political power, which stimulated popular opposition and its ultimate overthrow.

The Pythagoreans are credited with the arrangements of individual logical arguments in geometry, such as those attributed to Thales, into a logical sequence or chain of argument. They were regarded as having 'transformed the study of geometry into a liberal education', by developing the geometrical process itself, separated from its applications.

The Pythagoreans must therefore bear their share of the blame for the separation of pure and applied science, which has dogged civilization for centuries. Their basic ideas were adopted by Plato, and through him have had unfortunate as well as fruitful effects on the development of science.

Pythagoras and his followers worked out the substance of the first two books of Euclid. Their most famous contribution was Pythagoras's Theorem, a general proof that, in all right-angled triangles, the area of the square on the longest, diagonal side is equal to the sum of the areas of the squares on the two shorter sides. His proof was not the same as that given by Euclid, and there has been much speculation on the nature of the argument that he used.

The Pythagoreans discovered many relations between numbers arranged in columns and lines, as in the Mesopotamian tables of numbers. Their most famous numerical discovery is that the number corresponding to the square root of 2, which obviously exists, cannot be expressed in terms of the integers, 1, 2, 3 etc. Suppose, they said, that this number 1, which corresponds to the length of the diagonal of a square whose sides are one unit long, can be expressed as a fraction, in which the numerator and denominator are combinations of integers or whole numbers. Then it could be shown by a simple calculation that the denominator must at the same time be both an even and an odd number. This was impossible, so the number representing the length of the diagonal, or the square root of 2, could not be expressed in terms of the common numbers, 1, 2, 3 etc.

The Pythagoreans concluded that the square root of 2 must be fundamentally different from the common integers, and described it as 'irrational'. This discovery was both a glory and an embarrassment to the Pythagoreans. They had discussed that there was more than one kind of number, which was magnificent. However, as they supposed that the universe was made of numbers, some of which they now found were 'irrational', they concluded that the universe must be irrational. As a religious brotherhood, they found it awkward to believe that God had created an irrational universe, so they tried to keep their discovery of 'irrational' numbers secret.

Another of the great discoveries of the Pythagoreans was that the musical note emitted by a vibrating string depends on the length of the string. This revealed a numerical element in sound, music and art. From their point of view, this was a major confirmation that the universe is at bottom numerical. They had brought wave-motion, one of the most fundamental of phenomena, within the range of mathematics.

The discovery also stimulated fantastic speculations. Their followers supposed that the planets moved at distances from the sun
determined by the numerical relations between different musical
notes. This led to the idea that the solar system moves according to a
musical harmony. Plato said that this heavenly harmony, though
audible to God, was inaudible to humans. Two thousand years later,
the great Kepler wrote down what he believed to be the musical notes
of the heavenly harmony.

The systematic study of mathematical and scientific processes
promoted by Thales, Pythagoras and their followers initiated a
wonderful development during the next two hundred years. The idea
of a primary substance which goes through transformations led to the
idea of an element out of which all things are made. Some regarded
the primary substance as being continuous, like a viscous fluid;
things being made by convolutions and knots in it. Others thought of
it as cut up into uniform little units, or atoms; things were put together out of atoms, like a house out of bricks. The Pythagoreans
believed that numbers were atoms. For them, things were put
together out of numbers.

Democritus developed the idea of the atom, and the complementary notion of the void, or physical space; for if atoms, which are
separate, are the only matter there must be something between them
which has its own properties, even if these are only the properties of
empty space.

Besides formulating many of the most fundamental ideas which
still dominate modern thought, such as mathematical and scientific
proof, continuous fluids, wave-motions, atoms, void or space, and the
evolution of organisms, the Greeks carried the formal development of
some of them very far. This was particularly exemplified in Euclid's
magnificent development of geometry, and Archimedes's power and
subtlety in the solution of problems.

Archimedes elucidated the principles of stability in ships, and
calculated the volume of a sphere by methods that were within sight
of differential calculus. He showed his awareness of logical difficulties
in mathematics, which were not appreciated until within the last
hundred years. In his work on the determination of specific gravity he
gave a perfect demonstration of scientific method, and he handled the

application of mathematics to physical problems with an acuteness unequalled until the time of Newton.

Nevertheless, the Greek mathematicians and scientists had major failures. The Mesopotamian invention of a numerical system with place-value, which greatly increases the power of computation, was not appreciated by them. It was adopted or rediscovered and developed by the Indians, and introduced into common usage through the Arabian mathematicians.

The Greeks were more under the influence of the Egyptians. Their geometry, like the Pyramids, appealed to the visual imagination. They engaged in geometrical construction, which is the transference of the attitude of the architect into the realm of mathematical ideas. The Mesopotamians preferred 'calculations', rather than 'constructions'; they were more interested in commerce than in building.

The Greeks made great progress in astronomy, medicine and biology, besides mathematics and physics. Aristarchus, who was born at Samos in 310 BC, suggested that the earth and planets revolved round the sun. He calculated the size and distance of the moon and sun by correct principles, and to a remarkable degree of accuracy.

Hipparchus, born about 160 BC, gave the first mathematically acceptable general theory of the solar system. It was worked out in detail by Ptolemy at Alexandria about AD 150, and remained in use for more than a thousand years, until it was gradually superseded by the Copernican system.

Hippocrates, born at Cos about 460 BC, became the founder of a great medical tradition. He and his followers compiled a wide range of careful medical observations, which have survived more completely than other branches of Greek science. They exhibit a mature systematic study of disease, attributing it to natural causes, in striking contrast to the magical tradition of ancient medicine. The Hippocrateans incorporated many sayings that have passed into common speech, such as:

> Art is long and life is short.
> Desperate diseases need desperate remedies.
> One man's meat is another man's poison.

They also formulated the Hippocratic Oath, which has ever since been the expression of the moral principle that should govern the medical doctor's conduct.

The biological side of Greek science was particularly advanced by Aristotle. He used his influence as the former tutor of Alexander the Great to have extensive collections of zoological specimens made throughout his vast domains. This material was brought to Athens, where Aristotle organized its study and classification. This classification led him to classify types of arguments, and so to the development of the science of logic.

The Greeks excelled in the abstract sciences, but they did not entirely neglect mechanics. Their most notable mechanician was Hero of Alexandria, who probably lived about AD 60. Among the machines described by him are a simple form of reaction steam turbine and a coin-in-the-slot device, for automatically collecting money from visitors purchasing portions of holy water at the entrance to temples.

It is evident that the Greeks advanced and touched upon many of the ideas that are still in the forefront of science. Their development of the notion of mathematical proof strengthened the conception of cause and effect. Their views on fluids and atoms led to the notion of continuity and discontinuity. Refined versions of these underlie modern ideas on calculus, quantum theory, statistics and relativity. Nor was the notion of biological evolution absent from their speculations.

Finally, towards the end of their scientific development, about the second century AD, the Greeks began to adapt the chemical techniques evolved in metallurgy and medicine to the preparation of substances of special value. These techniques had been evolved over thousands of years in order to obtain metals and other useful things from raw materials known to furnish them. They were now adapted and developed for a different end: the direct making of wanted substances, rather than mere extraction from raw materials. Thus the old techniques were adopted by new magicians, who, instead of practising the old magical processes, developed the large-scale workshop techniques on a small scale. They miniaturized the old productive processes, because they were not interested in handling large quantities of materials, but in transforming small quantities of cheap ·

materials into rare things of great value. Their small-scale repetition of large-scale processes led to the creation of the research laboratory out of the productive workshop. These up-to-date magicians, both men and women, became active in Alexandria about AD 150; one of the most outstanding being Maria the Jewess. They set up small-scale distillation apparatus and flasks in research rooms, the recognizable ancestors of the chemical laboratories of today.

Development in China had important resemblances to, and equally important differences from, that in the West. It began with the settlement of the Yellow River valley by pastoral tribes, who gave up stock-breeding and became dependent on agriculture more quickly than the corresponding tribes in the valleys of the Indus, Euphrates and Nile. This was due to the monsoon climate, which made pastoral life unsafe.

The great dependence of Chinese development on agriculture is probably one of the causes of its stability and continuity, which provided the basis for its unsurpassed refinement in various directions. By the period 200 BC to AD 200 mathematics had progressed to the stage of accurate surveying, including the correct calculation of the areas of triangles and related shapes. They knew, and were able to prove, that the square on the longest side of a right-angled triangle was equal to the sum of the squares on the two shorter sides. They could calculate square and cube roots. They could calculate the volumes of pyramids and other building structures. The solution of equations by false position, from guesses above and below the correct value, had been discovered. It was introduced into Europe through the Arabian mathematicians as the Chinese Method.

About AD 263 the mathematician Liu Hui calculated the value of π to be 3·14159. The use of paper became widespread, and the wheelbarrow was invented. Chinese inscriptions on bone contain references to eclipses of the moon in 1361 BC, of the sun in 1216 BC and also to a nova, or new star. Chang Hêng in the second century AD used a water-mill for revolving a bronze celestial sphere, and invented a seismograph for recording earthquakes. Yu Hsi discovered the precession of the equinoxes, caused by the earth's top-like wobble arising from its spin, about the year AD 336.

In spite of these and many other advances, the Chinese did not develop generalized thinking about technical and scientific processes. There does not seem to have been a people in the Far East in a position to extract general principles from the rich achievements of thousands of years of Chinese civilization, in the way that the Greeks derived general principles from the geometrical and arithmetical discoveries of the Egyptians and Mesopotamians.

4. Why Greek science declined

The Greeks first appeared in history as fighting barbarians, primitive and uncivilized compared with the Egyptians and Mesopotamians, whose civilizations were already thousands of years old. Working on the data they found to hand, they developed Greek science with great speed from about 600 BC. Their progress in science was part of a general development in political ideas, economics, technique, armaments and military tactics. They were small in number, flexible, mobile and efficient; the commando force of a new civilization.

The Greek settlements on the Eastern Mediterranean islands and coasts, which had produced their local scientists, each with their own views, gradually came under centralized control. This was established first in Athens. The wealth and power of the city enabled it to become the centre of the new Greek scientific activity, the first assimilation of which into a consistent system was carried out there, by Plato and Aristotle. They lived only about two hundred years after the beginning of Greek science. They were still near to its original force, and reflected its greatness.

Nevertheless, retrograde as well as progressive tendencies already appeared in their work. The wealth and military success of Athens strengthened the aristocracy, and widened the division between the ruling and the skilled classes. The proportion of slaves in the population increased. The Greek social structure became more complex and stratified, and more like that of Egypt and the ancient civilizations. The intellectual interests of the rich and leisured grew in social status, while those of traders and craftsmen declined. Manual work became less reputable, and this affected the esteem of experimental science. Nevertheless, the positive influence of Plato and Aristotle was to prove very great. Plato inscribed over the entrance of his school, the Academy, 'Let no one ignorant of mathematics enter here.' His demand for mathematical descriptions of natural phenomena became a major influence in the creation of modern science.

Aristotle was Plato's most gifted pupil. His intellectual interests were more concrete, less purely theoretical, than his master's. He presently founded his own school, the Lyceum, where he could develop his own point of view independently.

Aristotle's study of animals exhibited an extraordinary range and acuteness of observation. Some of his observations of animal behaviour were not successfully confirmed until about a hundred years ago. He invented the basic terms for the description of animals, such as *species* and *genus*, and started the huge task of animal classification, to be continued by his successors.

But in spite of their great contributions, Plato and Aristotle were not *experimenters*. This side of Greek science was weak, and grew weaker.

The expansion of Greek power represented by the rise of Athens was carried to its furthest extent by Alexander the Great (356–323 BC), who absorbed Athens and conquered the Middle Eastern World. He ordered his commanders to collect material for his old tutor Aristotle, who thereby accumulated collections which were to form the nucleus of the first Greek museum and library, besides providing material for his personal researches.

Alexander decided to establish his world capital at his new city of Alexandria, which he founded in 331 BC. Eight years later, in 323 BC, he died at Babylon, at the age of thirty-three. Aristotle then left Athens owing to the tension following Alexander's sudden death, and died in the following year.

Alexander's empire was divided into three parts. One of these, under the rule of his general Ptolemy, was based on Alexandria. Ptolemy invited Strato, the director of Aristotle's school, the Lyceum, to organize an institution for promoting science and learning. Strato brought the Aristotelian library and collections to this institution, known as the Museum, which soon became the greatest centre of organized research. Euclid (330–275 BC) was one of its earliest professors, followed by Aristarchus, Archimedes, Appollonius, Eratosthenes, Hipparchus and many others, down to Hero (AD 60) and Ptolemy (AD 168).

Plato's Academy continued to exist at Athens for nearly a thousand years. The studies there became more literary and mystical. Its curriculum of mental rather than manual learning was easily adapted to the requirements of the increasingly non-scientific rulers of Europe.

The Romans occupied Athens in AD 168. Thenceforth, it became a centre where members of the Roman ruling class completed their higher education. It reflected Roman ideas. The Romans were conquerors, rulers and administrators; more interested in law and organization than in technique, except where it bore on warfare and government. They were interested in the improvement of weapons, the construction of roads, water-supply and military medicine. They invented the first hospitals for the treatment of the wounded in campaigns.

Their military conquests led to the capture of immense numbers of slaves, who formed a huge preponderance in the population. The control of the slaves was organized with the utmost thoroughness and brutality. Roman slavery was retrograde. The Egyptians and Mesopotamians were pioneers, and knew no other way. The Romans might have improved on the better methods of the Greeks, but failed to do so.

The obverse of this social development was the rise of mystical religions, of which Christianity was the most important. This provided the consolation of a happy life in the next world to those who were oppressed in this. The new Christian religion was more easily combined with Platonic philosophy than was the earlier Greek concrete and scientific outlook. This gave further life to the Athenian schools, and also contributed to the rise of Platonism in Alexandria.

Ptolemy, the last great Greek astronomer, who worked at Alexandria in the shadow of the growing mysticism, and whose system was to be used for more than a thousand years until modern times, was a less accurate observer of the stars than his predecessor and model, Hipparchus, three hundred years before.

Slavery prevented the Greeks, and still more the Romans, from correctly assessing the importance of the experimental part of science. They could not bring themselves to realize that the facts obtained with the aid of manual operations, despised in the ordinary affairs of life, were as necessary for science as pure thinking.

Slavery weakened the motives for the development of a science like the modern, in which theory and experiment are equally esteemed, and duly balanced.

5. The gestation of modern science

The Romans extended their empire on the basis of organized slavery, and in fact extended it beyond the limits that the backward technology of their slave society could safely support. A. H. M. Jones has attributed the fall of the Roman Empire primarily to its backward technology, which compelled the Romans to have too many hands engaged in production, and left too few for manning the army. Consequently, in the end the empire proved too weak to resist attacks from outside.

One of the greatest and latest attempts at renovation was made by Justinian (AD 483–565), who in 527 carried out a comprehensive reform of the law; he aimed at making it, and the imperial government, more efficient. Two years later, he closed the Athenian philosophical schools, which he regarded as decadent and subversive.

That year, AD 529, may be taken as the end of the era of Greek science, which from the time of Thales had lasted about nine hundred years.

Justinian's efforts were unavailing. The imperial organization disintegrated. It became impossible to trace escaped slaves and return them to their owners. The huge agricultural slave estates in Europe broke up into self-contained units ruled by local lords. These lords were forced to grant some rights to their labourers, in order to retain them in such chaotic conditions; they could no longer be kept merely as slaves. The bold and discontented ran away and lived as bandits, as there was no longer an efficient Roman army to hunt them down.

In the obscure Dark Ages that followed, efforts of organization on new technological bases were made. The ancient agricultural slave society gradually changed into the feudal system. From the Mediterranean to the North Sea, the slave was replaced by the tied

c

labourer, whose rights, though at first small, existed and gradually increased.

This movement to increase the status and repute of manual work was reinforced by the more clear-sighted of the leaders of the Church. In AD 543 Benedict founded the first order of monks in Europe. This was a disciplined organization, intended to combat the prevailing chaos by dedication to manual and intellectual work, as well as prayer.

Thus the construction of a new social order began in Europe. Life in it was very rough at first. Men were preoccupied with survival and security. Abstract Greek science was almost forgotten and unknown. It was irrelevant to contemporary needs. However, in the freer society superseding the Roman, technology began to advance. For example, the rough Anglo-Saxons in England made far more use of water-wheels for production than their more cultivated predecessors. While the old Roman Empire was undergoing fundamental internal changes, the new people on its frontiers became increasingly active, like the early Greeks on the confines of Egypt and Mesopotamia.

While the fierce Scandinavians and Germans were attacking from the North, a new people emerged from the harsh interior of Arabia, composed of the followers of the new prophet Mohammed (AD 570–632), a trader of Mecca. Slavery in the new Mohammedan society was less onerous than in the Roman, for in its early stages the new cult was anxious to secure converts; thus a slave in the old society might become a free man in the new, if he adopted the new faith. The Northmen and the Mohammedans swept into the failing empire with speed and success, the former over-running England and Northern France, and the latter capturing much of the Middle East and Northern Africa.

In a period of only about a hundred years the Mohammedans carried their conquests through all those regions, invaded Spain and then France. They were not halted until defeated at Tours in 732 by the feudal chieftain Charles Martel. Thereupon they retreated into Spain, which they continued to dominate for four hundred years, until 1136, when Cordoba, their cultural capital, was captured by Ferdinand III. They were ultimately expelled from Spain by Ferdinand V and Isabella, in the year in which their Italian explorer-captain, Christopher Columbus, discovered the New World of the Americas.

The Mohammedan threat to feudal Europe was a strong stimulus to its consolidation and development. In the eleventh century feudal Europe became strong enough to go over to the attack, and launched the Crusades into the Mohammedans' own territory.

The feats of the Northmen were equally spectacular. Within a period of about a century, they occupied Kiev, captured Constantinople and conquered Sicily. They invaded north-western France, learned the language and acquired the feudal traditions, together with new military techniques which had filtered through from the Mohammedans. Combining all these with their military prowess, they invaded England in 1066 and consolidated it as a single state. This early and permanent definition of the country was a great stimulus to its development, and an essential factor in its subsequent scientific achievement.

While the Europeans were engaged in reconstructing their social order, the Mohammedans were enjoying the fruits of their conquests. Their spiritual problems were solved for them by their new religion; the answers to all such questions were at hand. Secure in their military power and spiritual beliefs, they engaged in building splendid cities and studying the culture of their subjects. They were new people, they had no knowledge of their own and they read the works of all the conquered with open minds. To them, Greek, Latin, Indian and Chinese knowledge could be considered on an equal basis. As a consequence of this open-minded attitude, the Mohammedans became the founders of the internationalism which is one of the most striking features of science. They were eager and acute students. They excelled as critical encyclopaedists, of which Avicenna (980–1037) is the most famous.

The Mohammedans, from their desert ancestry, had a strong trading and travelling tradition. They were interested in such problems as the exact calculation of shares in goods and the allocation of family inheritances. One of the problems to which they devoted a great deal of attention was the exact calculation of the decline in value of a female slave with age, like a modern second-hand motor-car.

These concerns promoted numeral-mindedness, and they adopted the Indian numeral system. Their greatest mathematician, Al-Khwarizmi, was born at Khiva in Uzbekistan south of the Aral Sea in about 800. He became librarian to the caliph Al Mamun. After travels in Afghanistan and probably in India, he published in about 830 his treatise *Al-jabr wa'l muqabala*, from which the word *algebra* is derived. In it he showed how the quadratic and other equations, required to deal with the problems that have been mentioned, can be solved. *Al-jabr* was the manipulation by which a negative number was removed from the equation, and *muqabala* was the process of simplification by adding or subtracting equal quantities to both sides of the equation. Al-Khwarizmi dealt with five classes of quadratic equation. He called the unknown quantity that was to be found 'the root', after the root of a plant, which is hidden in the ground; and he used the word 'power' to describe the square of the root.

One of the most distinguished of Al-Khwarizmi's successors was Omar Khayyám, who was born at Nishapur in Persia in the eleventh century. He died in his native place in 1123. Omar Khayyám gave rules for the solution of three classes of cubic equation, and also for one bi-quadratic equation. He is also said to have asserted the famous proposition that the cube of any integer cannot be expressed as the sum of the cubes of two other integers. Khayyám's poetry is widely known to readers of English through the translation by Edward Fitzgerald.

The Mohammedans were devoted students of astronomy, to fix accurately the dates of their religious festivals rather than discover how the heavens worked; in this they resembled the ancient Mesopotamians. Their interest in the numerical relations between astronomical data led them to develop trigonometry. They compiled tables of sines, cosines, tangents, cotangents, secants and cosecants, and worked out the relations between them, using this material for the exact calculation of the times of prayers. They published treatises on the manufacture and use of the astronomical instruments still used in mosques for determining the times of prayers. The astronomical information was utilized by their sailors as a guide to navigation in the Indian Ocean. They recorded the existence of the Magellanic Clouds of stars, seen in the heavens of the Southern Hemisphere.

Mohammedan criticism of the science of their predecessors was assisted by the particular structure of their language, which is suited to analytical thinking. It led to notions of a philosophical language which would give an accurate description of phenomena, and, through Leibniz and his successors, to modern mathematical logic. An analytical criticism of Euclid by Nasir al-Din was the starting-point of the first attempt to formulate a non-Euclidean geometry, by G. Saccheri (1667–1733) in 1733.

The Romans, who were better at leading than learning, added comparatively little to the scientific and technical achievements of their predecessors. The Mohammedans, in contrast, modestly and energetically followed, revived and extended the achievements of the past. They repaired Cleopatra's canal in Egypt, and restored the ancient irrigation systems in the Middle East.

Of a still higher order of importance, they established a new scientific and technical development in Spain. They brought copies of their translations of the Greek mathematicians and scientists, and of their own critical and original works, to Cordoba, making it the most advanced intellectual centre of the age in Europe. Through Cordoba and southern Spain, the new feudal society of Europe began to learn the ancient Greek science, acquiring more knowledge through this channel than through the Crusades in the Near East. English scholars were in the forefront of this development. The monk Adelhard of Bath went to Cordoba in about 1120, disguised as a Mohammedan student. He returned with a copy of Euclid, the Latin translation of which served feudal Europe as a textbook for four centuries. Adelhard also introduced Al-Khwarizmi's trigonometry; and Robert of Chester, who studied at Toledo, translated Al-Khwarizmi's algebra in about 1245. Another of the outstanding translators was the Italian, Gerard of Cremona (1114–1187), who translated Ptolemy's *Almagest*, parts of Archimedes and Aristotle, and many of the Mohammedan scientists.

The Mohammedans introduced into Spain the engineering and agriculture that they had learned in the Middle East. They constructed immense irrigation works, and introduced the cultivation of sugar-cane and cotton into Europe. It was from this source that the Dutch learned the principles of hydraulic engineering which enabled them to start on their own development, which was to have exceptional significance for science. Through this source, too, the Spanish colonists of America acquired their knowledge of two of the crop plants, sugar cane and cotton, which were to have so great a role in American history.

The productivity of Moorish irrigated agriculture in southern Spain was so great that the annual income from it was larger than the combined incomes of all the monarchs in contemporary feudal Europe.

The Mohammedan genius for appreciating and extending the work of others is probably responsible for the transmission of the fundamental Chinese technical inventions. The harness of horses in the ancient world was grossly inefficient, owing to its being placed round the animal's neck. If the horse pulled hard it automatically choked itself; consequently, the tractive force of a horse was effectively little more than that of a man, and the availability of slaves removed the spur to inventive improvement. The Chinese had invented more efficient harnesses at least as early as AD 400, and modern ones by 1100. The knowledge of these more efficient harnesses reached feudal Europe. It was utilized to create a new cavalry and new cavalry tactics. The possession of improved military technique was one of the explanations why a few thousand Normans were able to ride triumphantly through Europe.

It is probable, too, that knowledge of the Chinese invention of gunpowder, the magnetic compass, printing and the escapement mechanism, which makes possible the construction of accurate mechanical clocks, reached feudal Europe through the Mohammedan channels of communication between Asia and Europe: caravan routes across the desert separating the Mediterranean and Eastern Asia, and sea routes along the coasts of Arabia and India.

Mohammedan scholars may have heard of the contents of Chinese textbooks of mathematics, published about AD 500, in which arithmetical and geometrical progressions were used to solve problems in weaving textiles. Other Chinese discoveries which may have reached them were methods of solving quadratic and cubic equations, published about AD 625. The Chinese determination of the value of π, published about AD 500, gave a result between 3·14115927 and

3·1415926. By 1247 the Chinese were printing negative numbers in black type and positive in red.

The escapement mechanism, by which machines are made to run at constant speed, was invented by Yi Hsing in AD 725; it was the crucial invention that made possible the subsequent construction of

3. Su Sung's clock escapement

accurate mechanical clocks. Knowledge of it seems to have reached Europe in the thirteenth century.

Chinese scientists were acquainted with the floating magnetic compass by AD 785, and knew that it did not point towards the true north. They used it in land-surveying. The invention of the compass appears to have been derived from the use of magnetic spoons by magicians, who claimed to foretell the future with them. The spoons were spun on polished plates, and divinations were made from the direction in which they came to rest. References to such spoons occur before AD 100.

Their seismograph was invented before AD 200. It consisted of a vase with a ring of holes around the rim, each lightly holding a

圖

漢 張 衡 候 風 地 動 儀 外 觀 之 想 像 圖

4. Chinese seismograph built by Chang Heng in AD 132

metal ball. When it was rocked by an earthquake the balls fell into metal receptacles, clanging them like a gong. The direction of the tremor could be deduced from which balls fell and which remained undisturbed. The Chinese scientists did not use their seismograph to measure the strength of the earth tremors. Their inadequate theoretical conceptions led them to believe that the strength of the shock was fortuitous, and therefore not worth measuring.

The fullness and care with which astronomical observations were recorded is illustrated by the Chinese observation of the great nova of AD 1054, of which no European mention has survived. It is the parent of the Crab Nebula, one of the most important objects in the sky for the development of radio astronomy and modern theories of the origin and development of the stars.

In spite of the great contributions by many peoples so far described, modern science had not yet been born.

Agriculture was invented in the Middle East and China, where the climate was warm and the soils sandy and light. These were easily cultivated. Scratching of the soil with wooden implements by unskilled slaves produced an adequate harvest. The wet clay soil of Europe defied such primitive technique. It demanded the invention of the iron plough, and a higher standard of individual skill than was obtainable from the slave. The iron plough gradually advanced the frontier of cultivable land across Europe, in the way that the frontier of cultivable land was advanced across North America during the nineteenth century. Population and wealth increased, churches and cathedrals were built, and learning was fostered. Great scholars appeared, such as Albert, Aquinas and Roger Bacon, who stimulated intellectual criticism, and created a receptive interest in abstract ideas and scientific experiments, knowledge of which was to be obtained from the translations of Greek works arriving through Mohammedan intermediaries.

Thus, in a social order that was at first barbarously inelegant compared with that of Cordoba and Baghdad, a new tradition of construction by increasingly independent labourers and craftsmen arose. Presently, their names began to be recorded, indicating that at last they were no longer anonymous slaves. The vital factor which, added to ancient science, made modern science possible was the social emancipation of craftsmen during the Dark Ages and feudal times. Once manual work acquired independent status, it became possible to give experimental operations the weight that they must have in a balanced science, consisting of a combination of theory and experiment.

Independent craftsmen first became conspicuous in the medieval cities of Italy, Flanders and Germany. They were a social product of the evolution of these cities, which grew around feudal castles. This new population, living outside the walls of the castle or bourg, became known as the *bourgeoisie*. Its interests diverged from those of the lord within the castle, and so it began to struggle for more independence. Within the bourgeoisie the interests of craftsmen diverged from those of merchants. The craftsmen protected themselves by forming guilds, while the merchants acquired leadership through their wealth.

One of the first important scientists to reflect this new European commercial spirit was Leonardo Fibonnaci, of the growing Italian city of Pisa, who was born about 1180. His father was employed in a customs house on the Barbary Coast, where, as a boy, Leonardo was taught arithmetic and Arabic. After returning to Pisa in 1202 he composed summaries and developments of the mathematics he had learned during his travels. Through him the Indian numerals were introduced into European commercial and practical life; an advance of vast importance, for without the facility which they provide in analysing the properties of material things, modern science could not have been developed. Leonardo dealt with parts of Greek mathematics, as well as Mohammedan algebra, and showed great power in solving problems. In 1225 he took part in a mathematical contest, conducted in the manner of a challenge at a feudal tournament. The contestants were asked to find a number, the square of which, when increased or diminished by five, would still remain a square. Leonardo gave the fraction $41/12$, which was correct. Then they were asked to solve a particular cubic equation by geometrical methods. Leonardo proved that this was impossible, but gave an arithmetical approximation, correct to nine places of decimals.

Roger Bacon (1214–1294) became the most eminent of the English medieval scientists. He became acquainted with the ideas of gunpowder, combinations of lenses to produce telescopes and microscopes, submarines, mechanically propelled ships and suspension bridges. He probably acquired most of these ideas from Arabic sources. He engaged in scientific experimentation, and reflected on its processes, expressing conceptions which foreshadowed modern experimental method.

He said that the only man he knew who was to be praised for his experimental science was Peter Peregrine of Maricourt. Peter wrote a treatise on magnetism in 1269. He explicitly pointed out the importance of manual skill in science. He investigated magnetism experimentally, and made a spherical lodestone which was in effect a model of the earth. He investigated the strength of attraction of a piece of iron at points on its surface, and found two where it was strongest, which he called poles.

Peter's work, no doubt inspired by information from Arabic sources, stimulated the European development of the mariner's compass, and provided William Gilbert with the basic material and method for his inauguration of modern magnetic and electrical science.

The monastic scholars studied the scientific literature arriving in Europe. They reflected on the views of Aristotle on motion, and began to advance beyond him.

About 1350 John Buridan denied Aristotle's belief that the velocity of a body is proportional to the force acting upon it, and inversely proportional to the resistance it meets. He contended that it was inconsistent with the fact that a stone does not stop immediately after it has left the thrower's hand. He explained the flight of the stone as due to an impetus which the thrower impressed into it; the impetus then carried the stone on, after it had left the thrower. Buridan used the notion of impetus to explain the bouncing of tennis balls, the vibration of bells and the free fall of bodies towards the earth. He was far from arriving at the notions of modern mechanics, but he helped to thaw the intellectual errors frozen for two thousand years in Aristotelian notions on mechanics.

The force behind this development was the expansion of agriculture, population and construction in Europe between 1000 and 1200. It was associated with an expansion of mechanical power, through the improved harnessing of horses, and the development of wind- and water-mills. In the absence of the slavery of ancient times, there was no other way of meeting the increased demand for power. The more efficient horse harnesses were utilized in pulling the iron plough, improved with wheel and mould-board to turn over the sod, through the heavy soil.

By the year 1000 water-wheels were common in Europe. Feudal lords set them up on their estates, and claimed a monopoly in them, extending their social power through ownership of the means of production. The water-mill was utilized for an increasing number of purposes: grinding and pounding products, such as ores and metals, besides milling grain. It was used for draining large areas of European marshland, stimulated by the increasing demand for cultivable land.

The windmill, mentioned by Hero of Alexandria in the first century AD, and probably originating in China, was developed by the Mohammedans, who brought it to Spain in the tenth century. It was adopted in Christian Europe in the twelfth century, and applied especially to flour-milling and water-lifting. The windmill helped to create whole new countries. A large part of Holland was made by pumping water out of the vast marshes at the mouths of the Rhine, largely by windmills. The Dutch have a national saying that 'God created the world, but the Dutch created Holland'. The reclamation of the extensive English Fen district, which made a substantial addition to England's cultivable land, was carried out by Dutch technique.

The development of wind and water machinery extended practical knowledge of force and mechanical motions. This was spread widely among increasingly independent craftsmen.

In Greco-Roman and Mohammedan society there had been sophisticated ruling classes which, within themselves, had been more advanced than those in early feudal Europe. There were families in Mohammedan society who devoted themselves to the cultivation of machines and mechanical devices, but as these were not much applied to industrial production, there was insufficient urge to work out their theoretical principles; they still depended too much on unemancipated labour.

The development of milling machinery led to the introduction of the lathe, which appeared before 1500 in south Germany. At about the same period water-power was applied to bellows for producing powerful blasts for furnaces; this enabled metals with high melting points to be cast. Mill machinery was adapted for the construction of the first European mechanical clocks, utilizing the escapement mechanism for regulating the speed, which was first invented by China. These early clocks became known during the thirteenth century. In the same century, the spinning-wheel began to displace the distaff for the spinning of wool.

Knowledge of the great Chinese inventions of gunpowder, the magnetic compass and printing spread to Europe. The Chinese invented gunpowder before AD 1000, and made cannon by the middle of the thirteenth century. These became known in Europe at the beginning of the fourteenth century. The manufacture of gunpowder from saltpetre and sulphur became a stimulus to industrial chemistry; the manufacture of cannon stimulated the development of metallurgy and mechanical engineering.

This development of technique on a wide front brought forth new types of men, who specialized in architecture and engineering. The separation of these from the directors of constructions and projects, as technical specialists, began in the thirteenth century. As Beaujouan has justly remarked, 'The growing interest in statics, dynamics, hydrostatics and magnetism must have gone hand in hand with the increased social status of craftsmen.'

The development of Northern Europe directed the attention of mariners more and more from the Mediterranean Sea to the Atlantic Ocean. Already by AD 1000 the Northmen had sailed to Iceland and had reached North America, which they attempted to settle. They did not succeed, because their general technique of living was not sufficiently advanced. Mohammedan seamen sailed down the western coast of Africa in the twelfth century. On their return, they sailed out into the Atlantic in order to catch the south-westerly winds, which carried them to Spain. These voyages were the chief cradle of ocean navigation. In the thirteenth century the magnetic compass came into use, and the ancient steering paddle was replaced by the hinged rudder, which made it more practicable to handle a ship under ocean conditions. The propulsion of ships without Norse or Mohammedan galley slaves demanded technical improvements in rigging and sailing.

The new technology, discovered and used by craftsmen of rising social status, was increasingly described in the language of its inventors. This was at first passed on verbally, and presently was written in the vernacular. It was a new literature, with an inspiration different from the accounts of ancient science in Latin, Arabic and

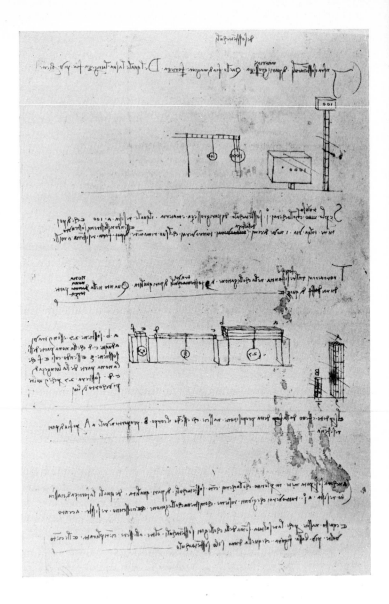

5. Leonardo's notes on strength of struts and loaded beams

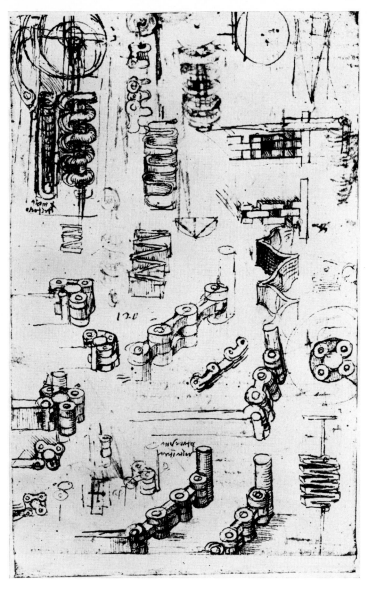

6. Leonardo's drawings of linked chains

Greek. Even where the subjects described were superficially similar to those in Greek science, the mental attitudes were so different that it was not easy to bring them to bear on each other.

The new independent craftsmen began to analyse the principles of their new crafts. They were the forerunners of Leonardo da Vinci, who knew little Latin and no Greek, and had no university education, but analysed the principles of painting and engineering from a scientific point of view. When he had conceived and entered into his scientific analysis of technical processes Leonardo sought elsewhere for relevant knowledge. He gained information from Archimedes's works, which assisted him to carry forward the lines of investigation that he had already begun. He became the supreme example of the emancipated craftsman whose combined manual and intellectual work in engineering, science, architecture, and painting achieved a prestige rivalling that of the mental philosophy of Plato and Aristotle.

Thus, as Beaujouan has said, the Renaissance, out of which modern science arose, 'though it acknowledged no masters other than those of classical antiquity, must, in fact, be regarded as the ungrateful daughter of the Middle Ages'.

6. The birth and development of modern science

Modern science was founded by the urban society that came into existence during the Renaissance, and first developed in the cities of Italy. Life in them was increasingly dominated by financiers, merchants and craftsmen, who advanced their various techniques. The increase in wealth had, among its various effects, two of major importance. The profits of trade and manufacture made men more intent on the improvement of the technical processes underlying them, and growing wealth gave more leisure to reflect on all processes, natural and artificial.

The effect of prosperity in the Italian cities was different from that in the cities of the ancient world, because of the differences between the various social classes. The new citizens of the Renaissance did not look at ancient science in the same way as its original creators. Their new life gave them an intense interest in the dead knowledge, but their attitude towards it was not the same as to the new knowledge which they were themselves creating.

At first their interest was antiquarian and literary. They excavated Greek and Roman ruins, and dug up statues and vases. They learned Greek, and sought for Greek manuscripts. The wealthy citizens became collectors who were concerned with the possession, rather than the content, of ancient rarities. They employed scholars and librarians to look after their collections and translate manuscripts. These wealthy patrons, who had a vital life of their own, were intensely curious about the kind of life that had been lived in the Greek and Roman past. The first Greek works to be translated were philosophical and literary, which threw light on how cultivated gentlemen thought and behaved in Greek times.

The Italian magnates imitated the social habits and literary tastes

D

of the ancient Greeks, but gave these a new content, because their own outlook and social ideas were different from those of the Greeks. They formed discussion circles, after the style of Plato's *Symposium*, and performed a kind of cultural charade. The superficial similarities between the art, architecture and literature of the Italian Renaissance and Ancient Greece concealed, however, their fundamental differences. Underneath the polite societies of each age there were two different social structures, and from this difference arose the different histories of science in Ancient Greece and science in the Renaissance.

When the patrons of the ancient learning had collected all the philosophical and literary works they could find, both in the original Greek and in Arabic, and had had them translated into Latin or Italian, they started on the heavier literature of mathematics and science. They were first interested in Greek scientific works primarily as collectors' treasures, and afterwards in their contents. They found that these had a bearing on their own activities as builders, seamen and merchants. They began to support the study of ancient science, to see whether it could provide information by which their wealth could be increased.

Growing coastal trade was an important factor in the prosperity of the Italian cities. Italian manufactures, such as fine textiles and glass, were exported from Genoa and Venice. The inhabitants of the Italian ports became interested in navigation and the building of ships. Christopher Columbus was born in Genoa in 1446, and Galileo studied the activities of the shipbuilders in Venice.

The exchanges of goods between Europe and Asia, mainly through Italy, required money operations based on exchanges of gold and silver. In consequence, there was a continuous influx of gold into Europe, and a growing export of European silver to the East. This provided the means by which the new Italian merchant princes, who ruled the cities and the countryside dominated by them, could patronize the poets and artists, and behave as they supposed Greek gods and heroes had behaved.

The export of silver gave a big stimulus to the development of metal mining in Europe. The rich silver mines in Bohemia were driven deeper, which raised severe problems of flooding and ventilation. These in turn caused engineers to improve pumps, and to study their mode of operation. This led on to the study of the properties of fluids in motion, both of water and air.

The discovery of new knowledge and the unearthing of the old stimulated education. No gentleman felt equipped to live in the new society without some acquaintance with the new learning. The Italian universities expanded to meet the need, and men of talent from the whole of Europe besides Italians flocked to the busy centres of the new knowledge. Many of the most talented of the students came from the outposts of Europe; Copernicus from the Baltic coast of Poland, Vesalius from Belgium and Harvey from England, to join in the bustle of study and research.

Copernicus was the scientist who made the biggest break with the past, and did more than any other individual to usher in the period of modern science. He was born at Torun on the river Vistula near the Baltic coast, in 1473. His father was a copper merchant and financier. When Copernicus was ten years old his father died, and his upbringing was undertaken by his uncle Lucas Watzelrode, who became Bishop of Varmia; this then contained a large part of Prussia, and the Bishop was virtually ruler of the country. He was an able ecclesiastic and statesman, who for long was more famous in Polish history than his nephew. He had studied at Cracow and Bologna, and he determined that his nephew should have similar advantages. He secured the appointment of Copernicus, at the age of twenty-four, before he had graduated, as a canon with business and management duties at his cathedral at Frauenburg. This gave Copernicus an income for life. Then he sent him to the University of Cracow in 1492, the year in which Columbus discovered America.

Copernicus began his studies at the start of the ferment caused by the greatest discovery of the age: the New World. The discovery of America and the subsequent circumnavigation of the world converted the idea of the sphericity of the earth from an intellectual deduction into a concrete reality. This made it easier to think of the earth as an individual object, separate from the heavens of fixed stars. Cracow was then the leading university in Northern Europe. Here Copernicus was taught mathematics by Brudzewski, the editor of the works of Purbach and Regiomontanus, the leading mathematicians and astronomers of the later Middle Ages. Greek was taught in the university, and so Copernicus was enabled to learn both mathematics and Greek.

After Cracow, Copernicus went to Bologna, ostensibly to improve his knowledge of law. He arrived there in 1496. He spent ten years in Italy, steeped in its cultural and scientific life, in the period when Cesar Borgia, Savonarola, Leonardo da Vinci, Michelangelo and Machiavelli were flourishing. There were thousands of students at Bologna. The city spent half its revenue on its university. Eminent professors were attracted by good salaries and fine buildings. The wealthy often settled in the city for years, pursuing the development of art, learning and science as a form of social distinction. Copernicus spent four years at Bologna, devoting himself more to mathematics, astronomy and Greek than to law. He became one of the first Polish scholars to master Greek. This was to prove of crucial importance in his astronomical researches, as he was able to read the Greek astronomers in the original, and not in faulty translations.

Copernicus's teacher at Bologna was Maria di Novara, a pupil of Regiomontanus. There was thus a direct connection between Copernicus and the most advanced of his immediate predecessors. Regiomontanus was a precocious German genius, who became astrologer to the Emperor Frederick III at the age of fifteen. With his teacher Purbach, he made a digest of Ptolemy's *Almagest*, in which trigonometrical functions were used extensively. Regiomontanus settled in Nuremberg, then the centre of the Renaissance in Germany. Art and mechanics flourished there, and especially manufacturers of clocks and scientific instruments. Outstanding among these was Behaim, who made navigational instruments used by Columbus and Vasco da Gama. An observatory was built for Regiomontanus, in which he improved the methods of astronomical observation, especially by more systematic attention to the correction of errors. He made more exact determinations of the times of observation, measured the position of planets by reference to the position of fixed stars and simplified astronomical calculations by more extensive use of trigonometry. Regiomontanus was the extreme example of one of the kinds of men who immediately preceded the appearance of the first modern scientists. Though his astrology and magic contributed much to his fame, they had a secondary place in his works; he had brought astronomy to a stage at which astrology and magic could be more easily jettisoned.

His pupil Novara took this development further. While making his living by astrology, Novara applied the improved methods of observation to the checking of the positions of all the stars recorded by Ptolemy. This led him to discover that the aspect of the heavens had changed since antiquity, an effect subsequently explained by Newton as due to the oscillation of the earth's axis, arising from the gyroscopic properties of the rotating earth. Novara was a Platonist and Pythagorean, believing that the explanation of phenomena was to be found in numerical relations.

Copernicus became one of Novara's collaborators. In 1497 they made an important observation on the eclipse of the star Aldebaran by the moon, which Copernicus subsequently used to prove his theory of the moon's motion.

Copernicus spent the year 1500, the Jubilee Year of Christianity, in Rome. Thousands of pilgrims came from all parts of the world, and he saw the city thronged with crowds of amazing men and women.

He returned to Frauenburg in 1501, without a degree in law. He was allowed to take up medicine, in order to make himself useful to the local population. His astronomical studies were considered a preliminary education for medicine. This arose from the doctrine of the microcosm and the macrocosm, according to which events in the microcosm, the human body, corresponded to events in the macrocosm, the heavens. Thus knowledge of the heavens threw light on events in the human body, and was supposed to be a guide to the causes of health and disease. As a doctor, Copernicus adopted old-fashioned methods. He believed in the efficacy of complicated pills, which were supposed to cure all ailments.

Copernicus now set off again for Italy, proceeding to Padua to pursue the study of medicine in its famous medical school. He went on to Ferrara, securing a legal degree from the Archbishop, who belonged to the Borgia family. On returning home after ten years of study in Italy, he was aged thirty-three, and qualified in law, medicine, mathematics and astronomy. He had also become a capable portrait painter. Frauenburg allowed Copernicus to become private secretary to his uncle, who lived in a palace ten miles from the cathedral. The political churchman treated him as a son, and apparently intended that he should be his successor.

Copernicus was largely free to follow his astronomical studies. He made observations for many years, gradually accumulating data for supporting his new ideas. He was not a particularly exact observer, but he could make observations which were good enough to decide between different theories. His reputation as an astronomer became international while he was still fairly young. In 1514, at the age of forty-one, he was called to Rome to advise on the continuing discussions on the reform of the calendar, a matter of major importance for settling the dates of church events, agriculture and the practical affairs of life.

Copernicus's first publication was not, however, on science. Like a typical Renaissance humanist, he made a Latin translation of a Greek literary writer, Theophylactus Simocatta. The book was published in 1509, with an introduction by one of Copernicus's former teachers, which contains the first published references to Copernicus's new ideas on astronomy.

Frauenburg owned one-third of the province of Varmia, and therefore the cathedral's administrative concerns were considerable. Copernicus was engaged in this administration. He was appointed governor of Allenstein Castle, and had to defend it against a siege by the Teutonic knights, which he did successfully. He was intimately concerned with the prosperity of the peasants on the Frauenburg estates. This led him to study the causes of the monetary inflation arising from the flood of American gold being brought by the Spaniards to Europe. He noted, following Aristophanes and preceding Gresham, that bad money drives out good. He held that bad money destroyed initiative, encouraged laziness and raised the cost of living. He regarded inflation, together with social discord, disease and poverty of soil, as the chief causes of the decline of nations. He advised that one mint should be established for the whole of Prussia.

Few scientists in history have enjoyed as wide an education and administrative experience as Copernicus. He was the reverse of the pedantic scholar. His ideas were rooted in the richest soil of the new Renaissance society.

He began to draft an account of his new astronomical ideas in 1536. Two crucial factors in his achievement were his knowledge of Greek and his profound identification with the new Renaissance life. Underlying his more realistic conception of the universe as a working machine was the increase in mechanical insight following on the increase in the use of machinery in industrial production. He embodied the complete exposition of his new theories in his great book, *Concerning the Revolutions of the Heavenly Spheres*, published in 1543, when he was on his deathbed. He had proceeded with the measured confidence of a man of experience, bearing the spirit of the new age. He had been in no hurry to publish, bringing out his masterpiece in his seventy-first year, at the very end of a busy life.

His studies of the Greek astronomers gave him a profound respect for their achievement. At the same time, he was a man of the Renaissance society, who respected the new achievements as much as those of the past. He had the self-confidence of the new social order to which he belonged. This helped him to recognize the differences between the revered observations of the Greeks and the equally admirable contemporary observations. He held that the new astronomical knowledge, accumulated during the last thousand years since the Greek effort had ended, was equally worthy of respect, and there was no doubt that it was not altogether compatible with Ptolemy, who had 'brought this science almost to its perfection'. A new principle was required to reconcile the old and the new observations.

Copernicus had probably heard from his teachers of the Greek theory that the earth revolves round the sun, and he made a search of the ancient literature to see what had been said of such ideas. He found references to them in Cicero, Plutarch, Heraclides and Ecphantus. Philolaus the Pythagorean held that the earth 'is moved about the element of fire in an oblique circle', while Heraclides and Ecphantus assigned a motion to the earth 'after the manner of a wheel being carried on its own axis'. Thus the notion of the revolution of the earth round the sun, and of its rotation on its axis, were discussed. From these suggestions Copernicus 'began to meditate upon the mobility of the earth'. He 'found at length by much and long observation' that 'if the motions of the other planets were added to the rotation of the earth and calculated as for the revolution of that planet, not only the phenomena of the others followed from this, but it also bound together both the order and magnitude of all the planets and the spheres and the heaven itself, that in no single part could one thing be altered without confusion among the other parts and in the universe'.

Copernicus dedicated his treatise to Pope Paul III, who had re-commenced the Inquisition. The question of the heresy of his theory did not seriously arise until about fifty years later. The Protestants were at first far more sharply opposed to it. Luther referred to Copernicus as a 'new astrologer who wanted to prove that the earth was moving and revolving, . . . such are the times we live in: he who wants to be clever must invent something all his own and what he makes up he naturally thinks is the best thing ever! This fool wants to turn the whole of astronomy upside down.'

In spite of this, Copernicus's first protégé came from Luther's Wittenberg. The twenty-five-year-old German professor of mathe-matics, Rheticus, journeyed to Frauenburg to learn his ideas. Copernicus, then sixty-six, was delighted with the intelligent young man, who devoted ten strenuous weeks to mastering the new theory. Rheticus wrote a summary of it in his *First Account*, which became the first published exposition of Copernicus's ideas. He compared Copernicus with Ptolemy, because the former, like the latter, had reconstructed the astronomy of his age. The Protestant philosopher Melanchthon was so annoyed with the *First Account* that he wrote that 'wise rulers should tame the unrestraint of men's minds'.

Rheticus urged Copernicus to complete his manuscript on the *Revolution of the Heavenly Spheres*, upon which he had already been engaged for more than thirty years. It consisted of a restatement of the contents of Ptolemy's *Almagest* on the basis of the principle that the earth revolves round the sun. Like Ptolemy, he assumed that the heavenly bodies moved in perfect circles, and he explained irregulari-ties in planetary motions by a corresponding assumption that the sun is not exactly at the centre of the planets' circular orbit but slightly eccentric. Nevertheless, by placing the sun at the centre, he showed that Ptolemy's eighty circular motions for explaining the heavenly movements could be reduced to thirty-four.

Copernicus's system did not give more accurate practical results than the Ptolemaic. His observations were twenty times less accurate than those of his successor Tycho Brahe, and he made many slips in

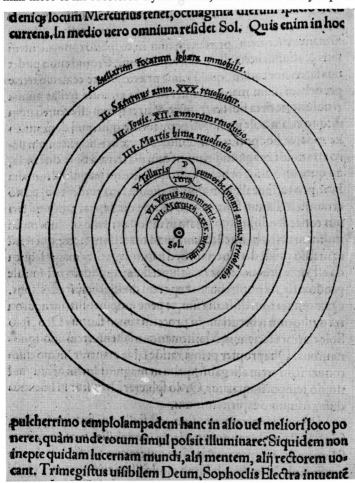

7. The solar system by Copernicus

his calculations. Profound thinkers were impressed by the simplicity of his system and its corresponding calculations, but practical men were slow to adopt it, because they were trained in the Ptolemaic system. Long experience had made them familiar with the latter, and Copernicus's system was not sufficiently perfect at first to give a decisive practical advantage.

The acceptance of the Copernican system involved a profound readjustment of man's conception of the universe. According to the ancient theory, the planets and stars were revolving around the stationary earth, and were not very far away. Man was at the centre of the universe, and the most important being in it. This belief had to be abandoned.

The idea of a moving earth required that the universe should be enormously spacious, in order to provide enough room for the earth to move in. Copernicus perceived the implications of the vastness of space, and he pointed out that the stars must be very distant, because their position did not appear to change when viewed from different points in the earth's orbit. He saw, too, that planets sweeping round the sun in a vast space needed some kind of force to keep them in their path; they could no longer be regarded as fixed on a revolving, transparent, solid sphere. He even hinted that this force might be found in the gravity which causes matter to fall towards the centre of the earth and cling together in spheres, like the droplets that coalesce to form a drop of water.

The replacement of the old, solid, compact, small universe by one of unending space, in which bodies were controlled not by rigid connections like rods but by physical forces, introduced a new order of flexibility and subtlety into thought about the nature and working of the cosmos.

Besides providing the soil in which modern physical explanations of the universe could grow, the undermining of the notion that earth and man were the centre of importance in the universe had momentous consequences. It was in conflict with current religious views; it put man in a more modest place in the scheme of things; and it destroyed the ancient microcosm and macrocosm theory, which for centuries had given astrology its apparent justification. The demonstration that in fact there was no close connection between events in the heavens and in the health and personal affairs of man, undermined astrology, and medicine was separated from astronomy. This had the further effect of separating biology from physics.

The denial of man's conceited claim that he and his earth were the centre around which the universe revolved made it possible to take an objective view of man, and thus provide a starting-point for such new sciences as the science of man, or anthropology.

A great deal had to be done before Copernicus's theory was completely and finally established. A body of more exact observations of the movements of the planets was required. This provided the data for the discovery that the true form of the planets' orbits was elliptical. Finally, there was the blunt ocular demonstration, through the invention of the astronomical telescope, of the existence of a system of moons revolving round the planet Jupiter, which provided a concrete pocket model of the solar system. About a hundred years were needed to collect this conclusive evidence, and a further fifty for its systematization to be completed by Isaac Newton.

The year 1543, in which Copernicus's *Revolutions of the Heavenly Spheres* was published, also saw the appearance of another great work, which opened the era of modern biology. This was Vesalius's *Fabric of the Human Body*. Andreas Vesalius was born in 1514, the son of Emperor Charles V's apothecary, who was a Belgian. His book, unlike that of Copernicus, was published early in his scientific life. Thus, in 1543, Copernicus was the old, while Vesalius was the young, protagonist of modern science.

Vesalius studied medicine first at Louvain and then in Paris. He was a wonderful student, working with tremendous speed and accuracy, and rapidly became a master of Galen's medicine, which had been the authoritative medical text for a thousand years. Vesalius had the typical self-confidence of the Renaissance man, which he combined with exceptional gifts of memory, observation and manual skill. He used his great talents to secure an appointment as physician to Charles V, and as soon as he had secured a high and well-paid position he virtually gave up scientific research. However, within the short period before this happened, he revolutionized anatomy.

He formed the plan of compiling a new treatise, to replace that of Galen. His work has a relation to Galen's which is similar to that of Copernicus's to Ptolemy's. In a comparable way, he reworked Galen's material from a new and independent point of view, drawing attention to Galen's errors with characteristic confidence and satisfaction, by putting his 'own hand to the business', as he expressed it.

His attitude annoyed the conservative professors at Louvain and Paris, and so he went to Padua, becoming professor there in 1537, when he was twenty-three. He objected to the old system of *demonstration* and *reading*, by which a *demonstrator* pointed out the features, and a *reader* read Galen or some other text to the students, while the professor sat above the class, expounding anatomy by words alone.

Besides carrying out a great deal of dissection himself, and classifying his material in a suggestive manner, Vesalius paid much attention to the illustration of his treatise. He secured first-rate artists to make the drawings. The plates of his *Fabric of the Human Body* exhibit a splendid combination of scientific and artistic quality. They established a new, modern, realistic standard in biological illustration.

Among the most important of Vesalius's observations was his careful record of his inability to discover any feature in the heart by which the blood could pass through the septum or wall, which divides the two halves or ventricles. Galen had said that the blood passed through pores in the septum, but of these Vesalius could find no trace. He continued to search for these pores, but when he produced the second edition of his book twelve years later he expressed his doubts of their existence still more firmly. Vesalius's observation, and his confident critical attitude based on thorough practical examination, formed the starting-point for the attack on the great problem of the circulation of the blood, which was the key to the investigation of the human and animal body as a working, functioning machine. Thus mechanical conceptions, being made more familiar by the increasing use of machinery in industry, entered biology.

The solution of the problem of the circulation of the blood, clearly posed by Vesalius, was found by William Harvey. He was born at Folkestone in 1578, the son of a merchant-adventurer in business with Venice and Constantinople. At the age of sixteen he was sent to Caius College, Cambridge, which had a leading reputation in England for medical studies. The college had been reformed by John Caius, who had studied under Vesalius at Padua and had probably lived in Vesalius's own house. Harvey graduated in arts and then, as the recipient of a medical scholarship, proceeded to Padua.

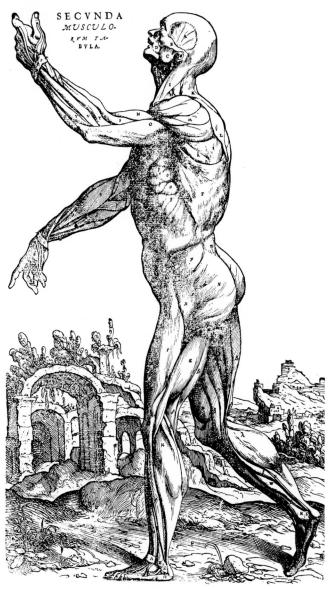

8. Vesalius's muscular system of a man

Copernicus, Vesalius and Harvey were all at various times students of Padua. This university was then the most radical in Europe. It was under the protection of Venice, the foremost anti-papal power. When Harvey arrived there about 1598 Galileo was expounding his work on mechanics and physics to large audiences, and questioning the principles of Aristotelian science. Vesalius had been succeeded by Fabrizzi, who continued anatomical research in the Vesalian tradition. He studied the veins especially, and published his work on *The Valves of the Veins* shortly after Harvey arrived. He also revived the study of embryology. Harvey became deeply attached to him and presently acted as one of his assistants. He followed him in both of his fields of research. Fabrizzi had noticed that the valves in the veins pointed towards the heart, and he invoked the principles of water-supply in trying to find an explanation of the blood's flow.

Harvey returned to England in 1602 with his Paduan doctor's degree. He started in medical practice in London and soon rose to the head of his profession. He married a daughter of Elizabeth I's physician, and himself became physician to James I, Charles I and Francis Bacon. He was always comfortably off, for his business-family managed his affairs for him. He lectured at the Royal College of Physicians, and pursued the lines of research to which he had been introduced at Padua. His lecture-notes for 1615, containing the evidence for the circulation of the blood, have survived. In them he said that the blood is constantly passed through the lungs into the artery which leaves the left side of the heart, as if driven 'by two clacks of a water-bellows to raise water'. He deduced from the effects of bandages on the arm that 'there is a transit of blood from the arteries to the veins. It is thus demonstrated that a perpetual motion of the blood in a circle is caused by the heart beat.' He had conceived the heart as a pump.

The final difficulty in proving the circulation of the blood arose from the fact that the transition of the blood from the arteries to the veins was through capillaries, which were so small that they could not be seen by the naked eye. Microscopes were not available, for they had not yet been invented. Harvey solved this problem by a masterly application of the mechanistic mode of thinking, which had been fostered by the increasing use of machinery in contemporary industrial production. Galen had vaguely conceived the movement of the blood as a gentle ebb and flow, analogous to the tide. He regarded the blood as sinking into the tissues like water percolating the soil, and subsequently rising as breath, like mists from the land. Galen had sought an analogy in the processes of nature; Harvey, belonging to the new age, sought it in machinery. Harvey had found that the blood did not ebb and flow, but circulated in one direction. Nor was it a gentle motion. He calculated the amount of blood pumped by the heart. As it beat about a thousand times in half an hour, and pumped about one-sixteenth of an ounce at a stroke, it would pump ten pounds five ounces of blood in half an hour, which passed in some way from the veins to the arteries. This was as much as the total amount of blood in the body. It could not be new-made blood supplied directly from the digestion of food, for such a large quantity could not be made by the body in that time. Consequently, roughly the same volume of blood must be pumped round and round the body in a continuous circulation, even if it could not be seen with the naked eye how blood passed from the arteries to the veins.

Though Harvey's work was so masterly and modern, it did not have much immediate effect on practical medicine. It was technically too far ahead of its time. Indeed, it had at first a retrograde effect on some medical practice, for it caused many doctors to give comparatively too much attention to the blood, and increased their belief in the efficacy of blood-letting.

While Harvey was meditating on the heart as a machine, Galileo was giving attention to the principles of the mechanical pump. Harvey may have acquired his grasp of these principles from Galileo's lectures. The modes of thought and scientific principles which were being invoked by Copernicus in astronomy and Harvey in biology were coming to the fore in other branches of science.

The demand for metals for guns, constructions and currency stimulated the development of mining, especially of pumps for raising water out of mine-workings. Outstanding descriptions of mine pumps and other aspects of mining were published by Agricola in 1556, in his great work *On Metals*. He was German by birth and, like Copernicus, Vesalius and Harvey, he went to Padua to study medicine; like them, he had wide cultural connections. He became a

A—Shaft. B—Bottom pump. C—First tank. D—Second pump. E—Second tank. F—Third pump. G—Trough. H—The iron set in the axle. I—First pump rod. K—Second pump rod. L—Third pump rod. M—First piston rod. N—Second piston rod. O—Third piston rod. P—Little axles. Q—"Claws."

9. A bank of pumps, Agricola

friend of Erasmus. He started a revision of Galen's medicine, and in 1527 was appointed physician in the mining town of Joachimsthal in Bohemia. The coins made from the silver of the local mines were called 'Joachimsthalers'. This was shortened to 'thaler', and the name was later adopted in America to describe the silver coin – the 'dollar'.

Besides giving excellent accounts of contemporary mineralogy, metal assaying, metallurgical chemistry, mining geology and prospecting methods, Agricola gave a comprehensive account of mining machinery, especially of mine pumps. He described seven kinds, including one which raised water 660 feet in three stages. Another consisted of multiple-stage suction pumps. He noted that a single-stage suction pump would not raise water more than 24 feet. Not far

10. Guericke's demonstration of the power of the vacuum

E

from Joachimsthal, at Magdeburg, Otto von Guericke followed up the development of pumps by inventing the air pump. He used it to demonstrate how great forces could be obtained from the pressure of the atmosphere. He drew upon the technical knowledge of the miners and of the Dutch engineers who drained Holland. The air pump was one of the most important inventions in science. It enabled accurate experiments to be made on gases, which are the simplest form of matter, and thus facilitated a rapid advance in physics, the science of the properties of matter.

The mineralogists, metallurgists and engineers were often far in advance of the academic scientists in their emancipation from astrology and magical ideas.

The Italian Vannoccio Biringuccio published a treatise in 1540 under the title *Pirotechnia*, in which he described the uses of fire in technological processes. He gave the first detailed account of the reverberatory furnace, in which the flames are directed on to the metal from above, and of the use of flame-colour for identifying chemical elements. He gave exact descriptions of many processes, such as the manufacture of gold and silver foils for making gold and silver thread, explaining how the cutting of the foils was done with very long scissors by women, 'who are much more patient than men' in operations requiring delicate skill. He described a gravity-fed, solid-fuel, heating stove, and gave detailed accounts of the complicated process of designing and manufacturing bells.

His style and intellectual attitude were as striking as his material. He was completely frank about what he knew and did not know. He showed an impatience of trade secrecy, and pointedly refused to claim any alchemical powers, though he indicated that alchemists might possess some knowledge which would be useful to technology. As his translator, C. S. Smith, has commented: 'inseparable from his account of an early stage in the growth of an experimental science is the picture of the beginning of capitalistic industrial economy as it related to a most vital type of production. Here we have science working hand in hand with industrial organization in beginning to produce a new society.'

The force which was undermining ancient astrology, alchemy and mysticism was the new social order, with its aim of exploiting the properties of matter. It was this which was enabling men to look at natural phenomena with a new realism, and was creating the conditions under which it became possible for Copernicus, Vesalius, Harvey and their successors to escape from ancient misconceptions, and thus found modern science.

The intellectual efforts of the new order were greatly helped by the publication of a third work of major importance in the crucial year of 1543. This was Tartaglia's Latin edition of Archimedes's works, which made the most acute mathematical and scientific mind of antiquity available to the new scientists, who had by then advanced through their own efforts to a point where they could begin to appreciate Archimedes's intellectual penetration. It was not possible for scientists simply to take up Archimedes's mathematics and science as he left them; they were a supreme product of the social order of his time. Nearly two thousand years passed before a new order arose on a different social basis that was sufficiently vigorous and sophisticated to equal, and exceed, the science and mathematics of ancient Greece.

Archimedes became available to the new European society when this had advanced to the stage when it could begin to understand and appreciate him. The progress of science does not depend merely on the construction of a chain of intellectual ideas, one link being added to another by men who happen to be exceptionally clever. It is an outcome of the whole life of the human society in which it grows, and cannot excel the basic values and virtues of that society.

7. Navigation, astronomy and physics

The discovery of America transferred the centre of the western world from the Mediterranean to the Atlantic Ocean. This created a pressure to develop ocean navigation in the Atlantic coastal countries; first in Portugal and Spain, and then in Britain and the countries along the coasts of the North Sea and the Baltic.

The Portuguese Prince Henry the Navigator, who lived from 1391 to 1460, founded an observatory on the southern coast of Portugal, where he promoted the application of astronomy to navigation, and encouraged the exploration of the Atlantic coast of Africa.

Navigation in the Mediterranean had gradually been built up on experience from reasonably full charts of the coasts and recorded knowledge of distances. In the North Sea and the Baltic the coasts were well known. Farther out, in the shallow waters of the continental shelf, depth-sounding with line and lead was a valuable aid. In the deep waters of the Atlantic, however, neither coast nor depth were known and could not be utilized. The navigator was forced to employ physics and astronomy. Attempts were made to use the magnetic compass. Columbus himself, however, discovered that the compass did not point constantly to the North, as he sailed from East to West, which made its use difficult.

The discovery of America changed fundamentally Britain's position in the world. From being a country on the fringe of civilization, she found herself on the future main line of communication. Her scientific interests and activities had hitherto been a minor and subordinate part of those of the continent of Europe, under the leadership of the Italians. Now the Italians found themselves on the fringe of the future development of Atlantic trade and of the New World, while Britain was placed between the Old World and the

New. The British changed their orientation from the East to the West, in science as in new possibilities of territorial acquisition and trade. They sought the solution of the problems of Atlantic navigation with a greater strength of purpose than they had ever applied to the scientific problems of continental Europe. The new situation enabled them to find themselves as a nation, and their exhilaration in achieving this was reflected in the cultural efflorescence of the Elizabethan age.

The British began by making fundamental improvements in methods of calculation, so that the intricacies of astronomical calculations required in ocean navigation could be facilitated and placed within the scope of sea captains and practical men. They developed the theory and practice of map-making, and created a new scientific-instrument industry to supply navigators with new types of astrolabes, quadrants and cross-staffs suitable for making observations at sea. The design and manufacture of magnetic compasses was developed.

The new science and technology were identified with practice. They posed problems to academic scientists, who left their universities to solve them and settled in London, the headquarters of finance and shipping, and of the trading companies formed to exploit the riches of the new-found countries and continents. Thus even when the scientists who created the new science and technology had themselves been educated in Oxford or Cambridge, they usually did their creative work in London and expressed the spirit of the City in their new work. They began to publish their books in the English language, instead of the Latin which had been customary for centuries, in order to make their contents accessible to the navigators and practical men who were usually not familiar with the ancient language.

One of the earliest of such scientists was Robert Record, a Welsh mathematician, who was born in 1510 and educated at Oxford. He published a treatise on arithmetic, with the title *The Ground of Artes*, in 1540. In this he used + and − signs. In his *Whetstone of Witte*, published in 1557, he introduced the use of the sign = for equality. This improvement in the symbolism of calculation was characteristic of the new development. It culminated in the invention of logarithms by the Scottish baron, John Napier, in 1594, when he was forty-four years old. It was the result of a direct attempt to reduce the complicated process of multiplication to the much simpler one of addition. Napier appears to have been the first to make a sound suggestion for the invention of a calculating machine. He did not, however, compile his logarithms in the form most useful to the practical man, that is, to the base 10. This huge task was undertaken by Henry Briggs, who was born in Yorkshire in 1561. Briggs was educated at Cambridge, and became first professor of geometry at Gresham College in the City of London in 1596. This was the first professorship of mathematics to be founded anywhere in England. Briggs journeyed to Edinburgh to meet Napier. When they met they stared at each other for fifteen minutes without speaking, lost in the profoundest mutual admiration.

The college which provided Briggs with a central position of influence was founded by the will of the financier, Sir Thomas Gresham. He was born in 1519 and became one of the richest men of his time. He was manager of Queen Elizabeth's finances. Gresham had been educated at Cambridge, and was keenly aware of the value of science and learning to England's growing industrial and mercantile society. He decided to leave his fortune for the endowment of a college in the City of London, where clerks and craftsmen, sea-captains and shipbuilders, mechanics and instrument-makers, and members of other growing trades and professions could be given the kind of instruction in geometry, astronomy, law, rhetoric, music and theology which they required as increasingly responsible and respectable citizens. What Oxford and Cambridge had been to the landowners, Gresham College was to be to the new industrial and financial society.

The mathematician Edmund Gunther was a colleague of Briggs, and a lecturer at Gresham. He introduced mechanical methods of using logarithms, while William Oughtred, another friend of Briggs, introduced the slide-rule in 1622. Oughtred, who was born at Eton in 1575, used the $\times$ sign for multiplication. John Wallis, Christopher Wren and Isaac Newton were among those who learned mathematics from his textbooks. The need for Gresham College as a centre for British science was illustrated by the limited influence of Thomas Harriot, friend of Walter Raleigh and Christopher Marlowe, who foreshadowed in his unpublished researches important advances in mathematics and astronomy. Among his innovations was the introduction of $>$ and $<$, as the 'greater than' and 'less than' signs in mathematics.

William Gilbert also was restricted by the private conditions in which he worked, before the era of Gresham College. Gilbert was born in 1540 and educated at Cambridge. He studied mathematics, among other subjects, and after graduating went abroad to acquire a medical degree. As a doctor he rose rapidly, becoming physician to Queen Elizabeth. Gilbert was a man of strong personality as well as fine intellect. He held his own at Court, and was himself a centre of intellectual interests. Elizabeth and her ministers were profoundly concerned with trade and warfare at sea. Every question and problem that maritime affairs raised was eagerly debated, and the attention of Gilbert's active scientific mind was drawn to them. He became interested in the use of the magnetic compass in navigation, and made

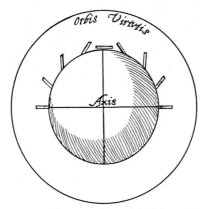

11. Gilbert's magnetic model of the earth

a comprehensive investigation of terrestrial magnetism in order to elucidate the scientific principles of compass navigation. He expounded his results in his treatise *On the Magnet and Magnetic Bodies*, the first major work on physical science by an Englishman and one of the chief contributions to the founding of modern science.

He elucidated the principles of magnetism by skilful experiments.

He followed Peter Peregrine in making a sphere out of a lodestone, to act as a model of the earth and its magnetism. He explored the properties of this model earth by means of a small compass which could be moved over its surface, as a magnetic compass is carried over the surface of the globe by a seaman in a boat. He compared the effects that he observed on his model with the reports that seamen brought

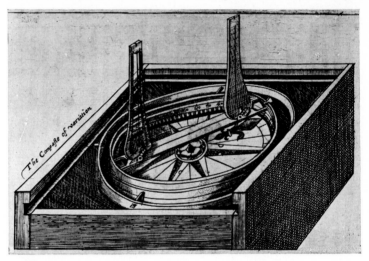

12. Barlow's compass

back from their oceanic voyages on the behaviour of the compass in various parts of the earth, and the changes that it registered in moving from place to place. He succeeded in explaining most of the effects observed by seamen. He diligently studied and acquainted himself with these men and their works. He refers to 'that most accomplished scholar Thomas Harriot, Robert Hughes, Edward Wright, Abraham Kendall, all Englishmen . . .' as having noted the differences of magnetic variation on long sea voyages. He referred to William Borough, William Barlow and Robert Norman as inventors and makers of magnetic instruments; the latter, indeed, had 'first discovered the dip of the magnetic needle'.

Gilbert's experimental investigation of magnetism led him to investigate the effects of electrification. He introduced the term 'electric' for describing the substances which could be electrified. It is from this word that 'electricity' is derived.

His study of magnetic and electric forces led him to speculate on the role of such forces in cosmology and the movement of the planets. The planets and the stars had, on the ancient Aristotelean theory, been supposed to be carried by rigid rolling spheres in which they were embedded. Gilbert's invocation of magnetism for this role deeply influenced both Galileo and Kepler.

Gilbert plainly ascribed the new experimental science of magnetism to the development of trade and industry: 'When, by the genius and labours of many workers, certain things needful for man's use and welfare were brought to light and made known.' Gilbert's great book *On the Magnet and Magnetic Bodies* was published in Latin in 1600. He came just too soon to be able to operate through the scientific organization which grew out of Gresham College, and the force of his genius consequently had less immediate effect in Britain than it might have had.

As Gilbert influenced Galileo and Kepler in physical science, so Napier and Briggs influenced Kepler in mathematics. Briggs convinced Kepler of the importance of logarithms, and Kepler's advocacy accelerated their rapid adoption in Europe. The British practical mathematicians were among the earliest to notice the Copernican theory.

The growth of science was rapid in other Atlantic countries. Simon Stevin in Holland, like the British scientists, made advances which were characteristic of the new practical and experimental science, and described them in Dutch, which he considered a particularly good language for the exposition of science. Stevin was born in 1548 in Antwerp, where he became a clerk in a counting house. He travelled in Europe, and afterwards had a post in the port of Antwerp. He then taught mathematics in Dutch to students of engineering at Leyden. Among his pupils was Prince Maurice of Nassau, who used advanced technique in his skilful military operations against the Spanish. Stevin became Prince Maurice's Quartermaster General and the organizing brain in his notable military campaigns.

Stevin's first publications arose out of his commercial experience. He published the first interest tables to be printed, as he was opposed on principle to secrecy in technique, a characteristically modern attitude. He was in favour of double-entry book-keeping. His most famous arithmetical innovation was his systematic use of decimals. In his book on the subject, published in 1585, he made clear for whom it was intended. He wrote: 'To Astronomers, Land-measurers, Measurers of Cloth, Gaugers, Stereometers in General, Money-counters, and to all Merchants, Simon Stevin wisheth health.' Parallel with his practical innovations, he advanced the theory of arithmetic. He based this on the notion of zero, instead of one, or the unit. He regarded zero as corresponding to a point in geometry. If a point corresponded to the concrete number zero, then a square root, which corresponded to a length on a line, was also a concrete number and was not an absurdity. This provided a clue to a consistent logical basis for algebra, which greatly facilitated its development.

As a port and military engineer, Stevin became concerned with mechanics and especially the principles of hydrostatics, an understanding of which was essential for the development of a country depending on a canal system of drainage and transport, that could also be converted into a system of military defences. Stevin mastered and extended the works of Archimedes on statics and hydrostatics, which had recently become more fully accessible, and which the new engineers were also becoming better qualified scientifically to appreciate. He proved that the pull of a body along the slope of an inclined plane was directly proportional to the steepness of the slope by appealing to a diagram, in which an endless string containing fourteen equal balls was hung round a wedge. The longest side of the wedge was horizontal, while one of the inclined sides was half as steep as the shorter side. Four balls rested on the longer side, while only two rested on the shorter. The chain of eight balls beneath rested in an even curve. Stevin appealed to the intuition that the string of balls would not slip round in a continuous motion, that is, to the intuition that perpetual motion is impossible. This solution is exceedingly ingenious. Stevin was so pleased with it that he made it the frontispiece of one of his books, with the legend in Dutch, *Wonder en is gheen wonder*, that is, 'the magic is not magical'.

Stevin clearly conceived the principle of the parallelogram of forces, necessary for the development of mechanics and scientific

DE
BEGHINSELEN
DER WEEGHCONST
BESCHREVEN DVER
SIMON STEVIN
van Brugghe.

TOT LEYDEN,
Inde Druckerye van Christoffel Plantijn,
By Françoys van Raphelinghen.
cIↃ. IↃ. LXXXVI.

13. Title-page with parallelogram of forces diagram, Stevin

methods of construction. In hydrostatics he proved that the pressure of water on the bottom of a vessel depends neither on the shape nor the volume of water, but only on the depth. From this he formulated the 'hydrostatic paradox', that the water or other fluid could exert a pressure on the bottom of the vessel which might be far greater than its weight. He deduced the pressure of water on the sides of ships, and proved that, for a ship to be stable, its centre of gravity must be below that of the water which it displaces, in addition to having a low centre of gravity as a whole. This is one of the principles of the scientific design of ships, and was a fundamental contribution to the new age of navigation and ocean trade.

Stevin also carried out the experiment on the rate of fall of weights, often attributed to Galileo at the Leaning Tower of Pisa. He and John Grotius dropped small and large balls of lead, and observed that they fell at apparently the same speed. They found, however, that a ball of thread fell faster than a single thread.

Stevin included among his achievements the harnessing of wind-power to land transportation. He made a carriage for Prince Maurice which carried twenty-eight people and was propelled by sails. It coasted along smooth beaches at a speed faster than a horse could gallop. After his fruitful life Stevin died at the Hague in 1620.

While Stevin was at work in the Netherlands Tycho Brahe was building, on the island of Hveen near Elsinore in Denmark, an observatory and research institute which he called Uraniborg, the 'city of the heavens'. In it he started the development of modern observational astronomy and collected the data necessary for further fundamental advances. His work was of a hard-headed and exact character, carrying the higher technical standards of the new social order into the ancient science of astronomy.

Tycho was born in 1546, two years before Stevin, at Helsingborg, on the other side of the channel from Elsinore, where Hamlet enacted the tragedy of his life. He died in Prague in 1601. Brahe's father was the governor of the castle at Helsingborg. It was intended that his son Tyge should be a statesman, and he was sent to Copenhagen University at the age of twelve to obtain an appropriate higher education. He studied rhetoric and philosophy, and became keenly interested in astrology which caused him to begin to learn astronomy. When he

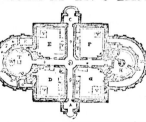

14. Uraniborg: 'The City of the Heavens'

was thirteen he witnessed a partial eclipse from Copenhagen. This excited a desire to learn more astronomy.

After three years of eager work on astronomy and mathematics, he was sent to Leipzig University, where he was supposed to pursue the study of law. He surreptitiously gave most of his time to his scientific interests, buying scientific books and instruments. He obtained tables of the motions of the planets, and discovered many indubitable mistakes in them. This was one of the crucial experiences in his life, for it impressed on him the need for more exact observation of the planets. He started systematic observation to this end before he was seventeen. From the beginning, Tycho exhibited a masterful directing spirit, besides great technical skill. He made his first important original observation, on the conjunction of Saturn and Jupiter, in August 1563, which was of intense interest to astrologers. The date compiled from the planetary tables then in use was wrong by amounts which varied from several days to a whole month.

Tycho then obtained a cross-staff, such as was used by navigators, for making observations. He found it subject to various errors and, not then being able to get a better one, he made a systematic record of its errors, so that future observations could be corrected for these errors. Tycho reflected the general tendency of the age in making astronomical observation more businesslike. Kepler regarded this event, in 1564, as the starting-point of modern astronomy, the science being newly restored in that year to its ancient status by 'that Phoenix of astronomers, Tycho . . .'

Tycho went abroad again to continue his studies, going, like Hamlet, to Wittenberg; where, also like Hamlet, he had dealings with the Rosencrantz and Guildenstern families, who were relatives of his. Wittenberg was then an active centre of astrology, astronomy and mathematics. It was for this reason that *Faust*, the imaginary figure of a magician, was described as having studied there. Tycho went on to Rostock, another centre of astrology and alchemy. Here he was involved in a duel, in which he lost part of his nose. He wore a plate of electron metal, an alloy of gold and silver, over the damaged place for the rest of his life. It added to the natural firmness of his expression, and made his appearance unmistakable. On his return to Denmark, the King assisted him to pursue his astronomical researches. He went abroad once more, to Augsburg, a centre of the new mechanical and instrument industry. He made use of these technical developments to secure the construction of very large and improved astronomical instruments. After his next return to Denmark he devoted himself at first mostly to alchemy, which was connected with astrology. Certain metals and certain planets were supposed to have similar influences in nature. For example, the planet Mars and iron were supposed to be so related, and the planet Mercury and the metal mercury.

Tycho's attention was finally fixed by an extraordinary event in 1572. While walking home one night in November from his alchemical laboratory, he suddenly became aware of a very bright star in the sky. It was in the constellation Cassiopeia, where he knew there had not previously been such a star. Before commenting on it, he asked other people whether they could see it, to prevent himself from being the victim of an illusion. As soon as he got home, he started observing it with a large new sextant, and kept it under observation for several months. He was unable to detect any movement of it relative to the fixed stars. It looked like an ordinary star, and it twinkled. It became so bright that it could be seen in broad daylight, and then after some weeks it grew dimmer, remaining visible altogether for a year and a half. He noted that its colour changed from white to yellow, and then to red.

There appeared to be no doubt that it was a new 'fixed' star. This was an entirely novel event in the history of European astronomy. The apparition of a new 'fixed' star was inexplicable in terms of the Aristotelian theory of the structure of the universe. The new star, which Tycho called the Nova, thus became one of the pieces of evidence that the Aristotelian theory could not be correct. Besides being of such importance to cosmology, the theory of the universe, the Nova has in itself proved to be a star of exceptional interest. It belongs to what is now called the 'super-nova' type. Its sudden flare-up is due to a nuclear explosion on a stellar scale; the explosion of something like a hydrogen bomb as big as a star. One of the most active radio sources detected in the sky in the mid-twentieth century has been in Cassiopeia. The violence of the explosion has caused the debris to move with such speed that it produces the radio waves observed by the radio-astronomers. Thus Tycho's star is retaining its exceptional importance to, and influence on, the progress of science.

Tycho wrote an account of the new star for his friends. At first they were incredulous, and then advised him to publish it. He objected to this on the ground that it was beneath a nobleman to write books, but he deferred to their persuasion when fantastic and erroneous accounts of the new star came in from many countries. Tycho's account, published before he was twenty-seven, became one of the chief marks of separation between ancient and modern science. His discovery that the apparently fixed part of the universe could change made it easier to question every feature of the heavens. He thirsted to discover whether there were any other new things in the supposedly fixed universe, which gave a tremendous stimulus to his natural aptitude for observation.

His reputation caused him to be invited to a professorship in Copenhagen University. He refused this at first, again because he regarded academic work as socially beneath him, but he finally accepted. He seems to have excused himself for lecturing in Danish on the ground that the Greeks were so good at geometry because they had studied the subject in their native language since their youth. He justified the study of astronomy on the basis of its utility for the measurement of time and its elevation of the mind. He held also that it was impossible to disbelieve in astrology without disbelieving in God; because man is made of the same elements as nature, his elements must be affected by the elements of heavenly bodies, as they have influence on one another.

Tycho planned and constructed on Hveen, with royal support, an institution which was more than an astronomical observatory. It had alchemical laboratories, craftsmen's workshops, a printing press, a library, a museum and guest rooms for visiting scientists. Feudal rights provided him with the means of living and an ample supply of servants. As an institution organized for scientific research, it probably influenced Francis Bacon's conception of organized science, described in *New Atlantis*.

Tycho's most important contribution was his development of systematic observation, with the best available equipment. He recognized that this could not be done without appropriate organization, staff and means. Hitherto astronomers had depended on occasional observations, which rarely revealed the subtler changes that became evident only after sustained and exact observation. His equipment was so extensive that he erected a second observatory near the main one, which he called the Stjerneborg, or 'city of the stars'. Some of the instruments in this were operated in cellars below ground, to shield them from the effects of wind and temperature changes. He kept up observations of the planets night after night, for twenty years, accumulating data on which a more advanced theory of the heavens could be based. He kept his records with scrupulous clarity and neatness. His body of observations remained unsurpassed in accuracy for a hundred years, until the time of John Flamsteed (1646–1719). He was the first to plot the orbit of a planet by taking observations throughout its course, instead of in a few positions. Consequently, he was the first to have determined the orbit of a planet entirely on an observational basis, and without any assumption of how it moved. This led him to the first adequately founded doubt that the orbits of the planets are circles. He suggested that they might be ovals like the contour of an egg.

Tycho's genius was not for theory. He had not the kind of mathematical imagination required to advance beyond the ancient fundamental conceptions, on the basis of his own observations. It was too much to expect him to have an equal genius in both theory and observation. However, he fully recognized that his observations had revolutionary implications, even if he could not fully work these out himself.

In 1577 a comet appeared, which he kept under systematic observation. He discovered that it was at a great distance from the earth, and could not possibly be an atmospheric phenomenon, as Aristotelian theory laid down. The comet of 1577 reinforced the implication of the new star of 1572, that the ancient conception of the universe, expounded with such completeness by Ptolemy, could not be correct. This caused him to view the Copernican system sympathetically. He admitted that it gave the correct mathematical results, but he could not accept it, because it seemed to him contrary to the laws of physics, besides being contrary to the Bible. He could

F

not believe that such a huge heavy body as the earth was in motion. He therefore suggested that the earth is indeed at rest and at the centre of the universe, with the sun, moon and 'fixed' stars revolving round it, while the other planets revolve round the sun. Tycho's theory was a practical man's compromise between the ancient and the Copernican theories. Copernicus's theory provided a difficult footing for further advance because it was so radical and also insufficiently precise.

In 1588 Tycho's patron King Frederick II of Denmark died. The new young king was less interested in his work, and reduced the royal subventions for Uraniborg. Tycho was not prepared to lower its standards and sought patronage elsewhere. He wrote a short account of his life and his astronomical instruments, with a summary of his discoveries. These included his compilation of accurate data on the position of one thousand stars, his immense collection of observations on the planets, the variability in the inclination of the moon's orbit, a fresh irregularity in the moon's motion, and more accurate data on the sun's motion. He printed this as a kind of prospectus, and dedicated it to Rudolph II, who indicated that he would be welcome in Prague, together with his equipment, and offered him the post of Imperial Mathematician.

8. Imperial mathematicians

Tycho decided to go to Prague. He arrived there in 1599, and was given a castle as a headquarters for his observatory. He set up his instruments and began his observations. He had difficulties, but one further supreme success. He succeeded in persuading the young German mathematician and astronomer, Johannes Kepler, to come to Prague.

Kepler arrived in 1600, when he was twenty-eight years old, and Tycho fifty-four. Kepler was engaged by the Emperor to compute new tables of the motions of the planets, from Tycho's observations. Tycho died shortly afterwards, in 1601. On his deathbed he begged Kepler to complete the tables, using his theory of the universe as the framework, in preference to that of Copernicus. Kepler completed and published the tables more than a quarter of a century later, in 1627; but he used the Copernican, not the Tychonian, theory. They are known as the Rudolphine Tables, in honour of their imperial patron.

Kepler was born near Stuttgart on December 27th, 1571. His father was a mercenary soldier, and his mother the daughter of an innkeeper. He was a delicate child, with weak eyesight, which hindered him from becoming an observational astronomer. His mother became a herbalist, and combined this with interest in magic and astrology. She was ultimately condemned for witchcraft, and was saved from being burnt at the stake only through a six-year legal battle by her son, who had become world-famous. With such a background, Kepler naturally grew up with an interest in astrology. Kepler's rearing was left to his grandparents, who sent him to a local school for craftsmen. It was about the only fortunate circumstance of his youth, for the Protestants in that part of Germany organized a particularly good system of schools in order to combat Catholic influence. His mental ability was immediately recognized, and at the age of seven he was transferred to the local grammar school, where gifted boys were encouraged by scholarships to qualify for the Protestant ministry. The prospect of such a career attracted Kepler. He had little difficulty in reaching the university, Tübingen, and graduating in philosophy, through his own efforts. There he attended the lectures of Mästlin, one of the best astronomers of the day, who taught the ancient Ptolemaic theory but in private expounded the Copernican theory of the universe.

Kepler's philosophical studies, and the humanist tradition of the age, interested him in Platonic philosophy. The explanation of the universe in terms of arithmetic and geometry appealed to his mathematical talent. Plato's theory that the planets emit heavenly harmonies profoundly appealed to Kepler. The discovery of proportions in the solar system, which would, as he believed, produce heavenly harmonies, was one of his strongest impulses to research. In one of his great works, the *Harmony of the World*, Kepler even wrote down in musical notation what he believed the heavenly harmony to be.

Kepler was engaged as a teacher of mathematics in the Protestant College at Graz in 1594. In addition to his duties as a teacher, he was appointed 'district mathematician', or astrologer. Throughout his life, Kepler obtained most of his income as an astrologer. The more experienced he became in astrology, the less he thought of it, and ultimately described it as the illegitimate daughter of astronomy, which nevertheless enabled her mother to secure a living.

Kepler had embraced the Copernican theory enthusiastically. It enabled the proportional distances between the planets to be calculated. This appealed to Kepler's Platonic ideas. He decided to search for 'the number, the size and the motion of the heavenly bodies', in order to discover 'why they were as they were, and not otherwise'. He exercised his wonderful imagination in conceiving different kinds of proportions between figures, and then comparing them with the observed planetary distances. He was amazed to discover that if a cube were inscribed in the orbit of Saturn, then the orbit of Jupiter would fit inside this cube. If a tetrahedron were inscribed in the orbit of Jupiter, then the orbit of Mars could be inscribed inside the tetrahedron.

He described the discovery in his *Mystery of the Universe,* which
had secured the attention of Tycho. Besides Tycho, he sent copies to

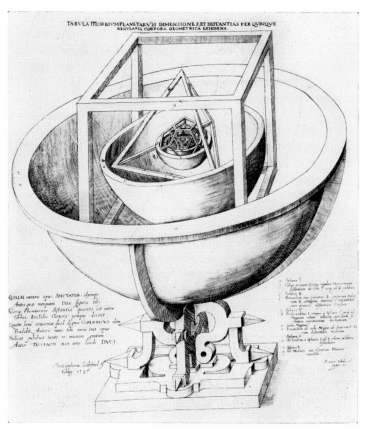

15. Kepler's picture of the solar system
as spheres in and around the five regular solids

Galileo and others. Galileo thanked him for his copy and con-
gratulated him on publicly supporting the Copernican theory,
which he was himself prevented from doing by circumstances. He
seems not to have read the book very thoroughly. Galileo's mind
was transparently clear, and Kepler's imaginings, consisting of

a mixture of fantasy and half-formed ideas of the profoundest genius, did not appeal to him, though he recognized his mental power.

Kepler said that geometry is a reflection of the mind of God. He believed that in discovering the numerical relations between the proportions of the solar system, he was discovering the geometrical skeleton upon which God had formed the universe. He regarded the geometrical figure of the sphere as a symbol of the Holy Trinity. The centre represented God; the surface the Son; and the volume the Holy Ghost. He dreamt about space-travel, and was one of the founders of science fiction.

Kepler's position at Graz became uncomfortable, owing to the pressure of Catholic power, which was in process of generating the Counter-Reformation. He decided to accept Tycho's proposal that he should move to Prague; he thought that the more exact data on the solar system which Tycho had collected would enable the discordances between the proportions of his system of inscribed figures and the solar system to be resolved. He had difficulty in getting on with Brahe, and after two years returned to Graz, where he tried to come to an understanding with the Catholic power. He drew up a prospectus of work which he was prepared to carry out under its patronage. In it he said that he proposed to explain the movements of the moon on the principle that its motion was non-uniform, and that there was a force in the earth which was the cause of the moon's motion. It followed from this theory that the farther the moon was from the earth, the slower it would move.

Kepler had begun to formulate an explanation of the solar system on the basis of physical forces. The ancient system, which explained the movements of the planets merely in terms of number and geometry, that is, in terms of kinematical theory, did not invoke physical forces.

He was not, however, able to come to terms with the Catholic power, and so he went back to Prague, where Rudolph appointed him Tycho's successor as Imperial Mathematician. The Emperor was far more interested in astrology and alchemy than in Catholic politics. He continued to reign until 1611, when the Catholic politicians, exasperated by his indifference, had him usurped by his brother. He died in Prague in 1612. Kepler remained in the city until after Rudolph died, and then moved to Linz.

The planet Mars presents the most pronounced irregularities in its movement. Tycho, who had kept it under very careful observation, asked Kepler to investigate his new data. One of the most difficult problems in the investigation arose from the irregularities in the motion of the earth itself. Thus the two sets of irregularities were mixed up and seemed inextricable. Kepler discovered how these two sets of irregularities could be separated, thereby greatly simplifying the analysis. It enabled him to consider the motion of Mars by itself. He computed what it would be, according to seventy different hypotheses. One of them gave a calculated orbit which agreed to within one-tenth of a degree with Tycho's observations. This would have been good enough for nearly all men, but not for Kepler. He knew that Tycho's observations were of a still higher accuracy than this. So, as Kepler put it: 'Since God has given us a most careful observer in Tycho Brahe . . . we should recognize this gift of God and make use of it. . . . But as they could not be neglected, these eight minutes alone have led the way towards the reformation of astronomy.'

He tried still more combinations of circular motions, but none of them gave sufficient agreement. Then, following the thought of Tycho, he tried egg-shaped ovals; finally, he tried the even form of oval, the ellipse. This also did not work if the sun were placed at the centre of the ellipse; but at last, by placing the sun at one of the foci, satisfactory agreement was obtained. The orbits of the planets were ellipses!

This was the end of the ancient dogma of the circle as the necessary, and only possible, form of motion for a planet. It was one more major marking point of the dividing line between ancient and modern science.

As the movement of planets in circles was evidently not a necessity or law of nature, their motion must be due to some other cause. Kepler began to suspect that it must be connected with the sun. He read Gilbert's book on *The Magnet*, and his notion that heavenly bodies might be influenced by magnetic forces. His reading of Gilbert helped to strengthen his belief that the sun influenced the movement of the planets by some kind of physical force.

He pursued his discovery that the planets move in ellipses, and by another effort of genius and immense perseverance, he discovered that the line joining the sun to a planet sweeps out equal areas for equal intervals of the planet's movement.

He published his first two laws of planetary motion in his *New Astronomy*, which appeared in 1609. In the following year Galileo announced his startling astronomical discoveries with his telescope. Kepler appreciated them with an unrestrained enthusiasm. He immediately began to think about the principles of telescopes. He invented the astronomical telescope, which gives an inverted image but bigger magnification, whereas Galileo's was of the opera-glass type, giving an upright image but less magnification. He worked out the geometrical theory of lenses, much in the form in which it is still presented in textbooks. This was accomplished during the turmoil at the end of Rudolph's reign.

At the same time he continued his pursuit of fundamental mathematical relations in the proportions of the universe. On May 15th, 1618 he discovered, after a vast number of trials and computations, that the squares of the times taken by two planets to describe their orbits are proportional to the cubes of their mean distances from the sun. This third law of planetary motion was indeed a wonderful discovery, and Kepler himself triumphantly said so. Shortly after making it, he wrote:

> 'I give myself up to sacred frenzy. I scornfully defy mortals with the open avowal: I have plundered the golden vessels of the Egyptians to furnish a sacred tabernacle with them for my God, far from the borders of Egypt. If you forgive me, I shall be pleased. If you are angry with me, I shall bear it. Well, then, I cast the die and write a book for the present, or for posterity. It is all the same to me. It may wait a hundred years for its reader, as God too has waited six thousand years for a spectator.'

In addition to his planetary laws, he contributed to many other parts of astronomy. He ascribed the tides to physical forces from the moon. He held that the sun's corona, seen during solar eclipses, was part of the solar atmosphere, and he explained the behaviour of the tails of comets, which point away from the sun, as due to a solar repulsive force. Besides his physical optics, he promoted the use of logarithms, and in response to requests to calculate the volume of casks with their curved sides, he advanced towards the invention of the calculus.

The richest contributions of Kepler's genius arose from the fertility of the unconscious part of his mind. He conjured up extraordinary ideas from its depths. In contrast, his great and slightly older contemporary, Galileo, had a mind which worked primarily with conscious thoughts. Galileo was transparently clear and logical, and in comparison with Kepler appears more rationalistic and modern.

9. The last great achievements of Renaissance science

Galileo Galilei was born in Pisa on February 15th, 1564, in the same year as Shakespeare. He died in 1642, the year in which Isaac Newton was born. He was descended from one of the leading families in Florence. His father was a distinguished musician, whose works were studied by Kepler when he was trying to discover the harmonies in the heavens. He was an outspoken supporter of free intellectual enquiry, which probably had an important influence in forming Galileo's attitude. The family was not, however, prosperous. Galileo was sent to a Jesuit school near Florence when he was twelve. He had a lively mind and strong memory, which enabled him to recite long stretches of poetry. His earliest considerable lecture was a piece of literary criticism, in which he discussed the place and size of Dante's *Inferno*.

His father saw that he was better suited for a scientific profession than for business, and so he sent him, at the age of seventeen, to study medicine at Pisa. There Galileo's professor was the eminent physician and botanist Cesalpino. He attended the lectures on Aristotle, of which he made careful note. He greatly respected Aristotle but questioned his ideas, in his father's spirit of free enquiry. His disputatiousness, acuity and great intellectual energy earned him the nickname of 'Wrangler'.

Soon after entering the university, when he was sitting in the chapel during a service, a swinging lamp attracted his attention; he began to watch it, and gained the impression that the time of the swing was independent of its size. When he got home he checked the impression with a bullet and a piece of string. He was eighteen when he discovered the property of the pendulum which was to make it so important in the development of the clock.

Galileo did not become interested in mathematics until his second year, when he happened to see the mathematician Ricci giving the pages of the Grand Duke of Florence a lesson in Euclid. He suddenly saw its significance almost instantaneously. His acquaintance with geometry and Ricci led him to the study of Archimedes, whose works first opened to him the full power and meaning of science. He learned from Archimedes how to use mathematics to make physical experiments give more precise and deeper information. Galileo adapted Archimedes's method to modern problems. He thereby became the first exponent of modern scientific method in a form with which scientists of today feel themselves at home. Perhaps Galileo's greatest achievement was to make scientific method more explicit.

Galileo's work on the pendulum, and accurate experimental determinations of the specific gravities of substances, after the style of Archimedes, attracted attention. On the one hand, his mind was clarified by Archimedean logic, and on the other, the accumulated experience of the emancipated and developed crafts helped him to gain an increased insight into how bodies actually behave.

He had, however, no academic appointment, as he had left Pisa University without a degree. He earned something by coaching, and his friends tried to secure him a professorship. He was turned down by five universities. Fortunately, the mathematics chair at Pisa became vacant in 1589, and Galileo was appointed to it. He had now to teach Aristotelian science as part of his profession. He therefore made a systematic investigation of Aristotelian mechanics, and the additions made to it by the medieval Aristotelians.

The invention of gunnery and development of machinery had made a precise understanding of the behaviour of rapidly moving bodies, especially of bodies falling freely, like cannon balls, of great practical importance. The difficulty of finding out exactly how freely falling bodies behave is that they fall so fast. Instrument-making was not yet sufficiently advanced to do this directly. Galileo evaded this difficulty by slowing down the fall, but not changing its character. He did this by rolling small metal balls down an inclined plane, assuming that they would follow the same law of fall as if they had been dropped vertically, but at a slower speed.

He obtained a smooth beam of wood about eighteen feet long, and made a groove along the top edge. Then he propped one end about one to three feet higher than the other. He rolled small smooth metal balls down the groove, and they ran sufficiently slowly to be measured with reasonable accuracy by the means at his disposal. He measured the times by means of a water-clock, opening and closing the spout with his finger as the ball passed the beginning and the end of the stretch on the groove. He said that when a ball was repeatedly rolled a certain distance down the groove the measurements of the time taken did not vary among themselves by more than 'one-tenth of a pulse-beat'. From his analysis of the way in which the speed of the ball increased, he obtained an experimental proof of the law of acceleration under gravity, and a precise measurement of the rate of acceleration.

He considered what would happen when a ball was given a push up the groove. If the inclination of the beam was very small the speed of the ball would decrease very slowly. If the beam were level and without friction, then the ball would go on for ever, without losing any of its original speed. Thus a body would continue in its state of motion unless interfered with; this contained the notion of inertia.

He showed that the motion of a particle projected out of the vertical, like a cannon ball, could be resolved into two velocities: one along the vertical, and the other along the horizontal plane. These could be represented on a graph. The track of a cannon ball, if free from the resistance of the air, would, he pointed out, in fact be a parabola, for its velocity along the horizontal would be constant, whereas its vertical velocity would increase as the square of the time of fall.

Galileo was appointed to Padua in 1592, where he was given a modest salary but an excellent intellectual reception. He remained there for eighteen years, lecturing to large audiences, and pursuing many-sided and fertile researches. He invented his sector for simplifying calculations. It consisted of two inclined rulers hinged at one end, so that they could be moved over a quadrant. The rulers and the quadrant contained markings which enabled various types of calculations to be made, such as rates of interest, the extraction of roots and the volume of solids (for example, embankments in fortifications). A large demand arose for this instrument, which has ever since been part of the equipment of engineers.

He attracted students from many parts of Europe. Among them was Ferdinand, a future Emperor of Germany. Galileo lived in a large house, in which he accommodated about twenty students, with a garden where he liked to discuss science with his pupils, while he dug and pruned, or took supper under the trees.

In 1604 another nova, or new star, appeared, which had a similar effect on Galileo as the nova of 1572 had on Tycho. It stimulated his interest in astronomy. Its incompatibility with the ancient idea of a system of fixed stars increased his conviction of the truth of the Copernican theory. With this patent new evidence, and in the freer atmosphere of Padua, he now felt able to support the Copernican theory in public. Venice was then powerful enough to prevent Rome from interfering in intellectual matters in her territories.

Galileo had in the meantime come in contact with the Grand Duke of Tuscany. He was engaged as holiday tutor to his son Cosimo Medici, then a boy of eleven.

In 1609 Galileo heard of the Dutch invention of the telescope. He immediately made one of his own and turned it on various objects with astounding results. The Venetian rulers ascended the famous Campanile tower and saw far-off shipping brought apparently near. They recognized the military and commercial value of the invention at once, raised his salary and confirmed his chair for life. Galileo made a bigger telescope which would magnify thirty times and turned it on the sky. In effect, he opened a window on the universe, for a marvellous series of discoveries was revealed. The Milky Way was seen to consist of myriads of separate stars. Mountains were perceived on the moon, and their height estimated in miles from the lengths of their shadows. He saw the sphere of the planet Jupiter, encircled by four moons.

Galileo quickly wrote an account of the spate of discoveries, with the title *The Messenger of the Stars*, or Sidereal Messenger. It was simple, descriptive and vivid, and caused excitement far beyond the frontiers of the scientific world, to be compared only with the effects of such modern discoveries as the release of atomic energy. The nature of his telescopic discoveries was quite different from that of his foundation of mechanics, which could appeal at that time only to a few advanced experts. He provided every man, in addition to the

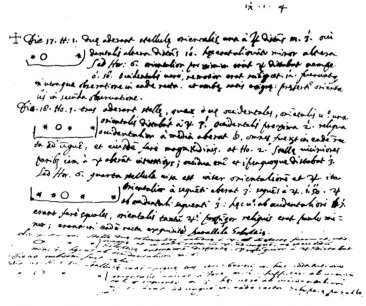

16. Page and diagram from Galileo's manuscript, the Sidereal Messenger

small number of scientists, with an immense extension of qualitative facts which could be appreciated without mathematical specialisms.

The observation of Jupiter and its four revolving moons was particularly important. If God had made a model Copernican system, might He not have made the solar system on the same plan? The spread of this view among the public did more to secure the acceptance of the Copernican system than the recondite mathematical arguments that appealed to philosophers. The existence of Jupiter with its moons was not a logically conclusive proof, but it was more convincing than logic.

Galileo now used the great fame that he had acquired to obtain a sinecure in his native land. He informed his old pupil, now Grand Duke Cosimo II of Tuscany, that he desired to write several treatises on his discoveries, in particular on astronomy and on mechanics. He wished to find a well-paid post which freed him from the chores of university lecturing, so that he could devote himself entirely to research and writing. Such a post was created for him, with the title of First Mathematician of the University of Pisa, at a high salary but with no teaching duties. Galileo's friends advised him not to take it, for they foresaw that the Duke of Tuscany would not be able to provide him with the intellectual protection that he had enjoyed at Padua under the wing of Venice. The Duke deeply admired Galileo, but his position depended politically on the goodwill of Rome. Ultimately, because of this dependency, he had to do what Rome wanted.

At first everything seemed to go splendidly. Shortly after settling in Florence, Galileo discovered the phases of Venus, and pointed out that these were a further confirmation of the Copernican theory. He observed the sun-spots, and deduced from them that the sun is rotating. He made further discoveries about the moon, and prosecuted researches in hydrostatics. He exulted over his critics, who became increasingly exasperated. The Jesuits especially were annoyed, for one of their own body, Scheiner, had previously observed the sun-spots, but as these are not mentioned in Aristotle, Scheiner had not been allowed to publish his observation.

Galileo's pro-Copernican opinions were now systematically attacked as contrary to theology. He confidently undertook to argue that this was not so. He was prepared to explain theology to the theologians. He believed that Duke Cosimo would see that he came to no harm. He went to Rome in 1616, sure that he could persuade the Pope, the Cardinals and the Inquisition that his views were correct. He was received with great honour, but he did not apparently perceive that, whatever his intellectual acclaim, he was having no political success. The Grand Duke's ambassador in Rome was alarmed by Galileo's behaviour. Galileo did not seem to understand that his opponents believed him to be undermining the authority of the Church, of which he protested himself a loyal member.

While he thought himself to be making great progress with his persuasion, he was astonished at being called upon by the Inquisition to disclaim his belief in Copernican doctrines, which he did. He returned to Florence humiliated. He wrote a pamphlet in which he criticized the views of Jesuit astronomers on comets. In it he expressed the opinion that 'motion is the cause of heat', and distinguished between such properties of bodies as size, shape and quantity, and those revealed to the senses, such as smells, tastes and sounds, which he regarded as subjective; this was the distinction between primary and secondary qualities, which has held a major place in modern philosophy. This pamphlet, with the title of *The Assayer*, enraged the Jesuits. He visited Rome again in 1624 and was loaded with presents, but his views were not accepted. He worked at his *Dialogues concerning Two World Systems*, believing that this would finally bring conviction. The manuscript was sent to Rome to be read, and certain emendations were recommended, one of which included the Pope's own argument against Galileo's theory of the tides. Galileo incorporated it, and the work was duly published, in 1632.

Then it was found that Galileo had dealt with the Pope's argument ironically, putting it into the mouth of the simpleton in his dialogue. The Roman authorities became very angry, believing that they had been tricked and insulted. The sale of the book was immediately stopped. Galileo was summoned to Rome for examination by the Inquisition. After long investigations he was forced to recant on his knees, under threat of torture, his belief in Copernicanism. With 'sincere heart and unfeigned faith' he said that he did 'adjure, curse and detest the aforesaid errors and heresies'. The story that he muttered, 'none the less, it moves', is unfounded.

Galileo lived under house-arrest for the rest of his days. He completed his second great work, on *Two New Sciences*, and smuggled it into Holland for publication, where it appeared in 1638. Even in his last years he was making discoveries. He observed the libration of the moon, that is, the slight variation in the moon's face. Newton subsequently showed that it arose from irregularities in the moon's motion. In 1637 he mentioned that the period of a pendulum is proportional to the square root of the length of its thread, and in 1641, at the age of seventy-seven, in the year before he died, he experimented with the pendulum for regulating clocks. His investigation of the properties of liquids led him to recognize limitations on the ancient

17. Galileo's design for a pendulum clock

G

theory of nature's horror of a vacuum. He pointed out that as a suction pump would raise water not more than about twenty-four feet, nature's horror of a vacuum was limited to about twenty-four feet of water. His pupil Torricelli extended his researches, and two years after Galileo's death invented the barometer with a vacuum above a column of liquid.

Galileo's fertility of discovery and intellectual energy were almost unparalleled. He illustrated, too, in his personality some of the characteristics of modern scientists. He was apt to believe that because he spoke with authority in physical science, he could handle on equal terms arguments in other branches of knowledge, such as theology and politics. For one whose logic was so penetrating in science, he seemed to be short-sighted in other directions. Galileo was the product of a declining, as well as the creator of a new, age. While his intellect shone, his personal life reflected the lack of dignity and the contradictions of a social order in dissolution.

10. The English explosion

England was one of the countries where the new social forms and scientific life were developing most quickly. In 1641, the year before Galileo died, political power had been wrested from the ancient monarchy, clinging to its divine right, by the merchants of London and the more business-minded landowners, represented by the Parliamentarians. Within a few years, the English social order was profoundly changed. An atmosphere of combined rationalism and enthusiasm arose, different from the decadent glories of Italy. In it, trade and science flourished wonderfully.

The scientific aims of the new age were splendidly foreshadowed in the works of Francis Bacon, who was born in 1561 and died in 1626. He extracted from the history of science, in recent and ancient times, a conception of scientific method, in which observation, classification and experiment were to lead to the formation of theories. These in turn would lead to more penetrating experiments and still deeper theories, until knowledge had been vastly extended, even to the effecting of immortality, 'if it were possible'. He proposed 'a total reconstruction of sciences, arts and all human knowledge', to extend the dominion of the human race over the universe. He foresaw the possibility and needs of the space-age. He did not restrict the application of scientific method to physical problems; it was to apply also to 'mental operations, logic, ethics and politics'. All the phenomena of the universe were to be investigated, according to a planned programme.

In his *New Atlantis* he sketched a new form of social organization, controlled by a scientific society in the interests of progress and human welfare. He had proposed in his projected work, *The Great Instauration*, to outline how human society could be reconstructed on scientific lines, with unending prospects of welfare, discovery and power; but he was not able to complete more than parts of it, represented by his *Advancement of Learning* and his *Novum Organum*, or New Method.

The triumphant Parliamentarians and their intellectual supporters attempted to carry out Bacon's ideas. Foremost among them was John Wilkins, who became a brother-in-law of Oliver Cromwell. He was born in 1614. In 1638 he published a book on *The Discovery of a New World*, and in 1641 a *Discourse concerning a New Planet*. The first contained arguments for supposing that the moon was a habitable world. It showed him well read in medieval and contemporary science, including Kepler and Galileo. His discussion of the nature of the moon's surface was not unlike that of modern space scientists, trying to foresee in what sort of conditions travellers to the moon will have to land. He argued that, as there was then no Drake or Columbus to make such a voyage, 'may not succeeding times raise up some spirits as eminent for new attempts, and strange inventions, as any that were before them?' He discussed the equipment that the space traveller would require to survive. He thought that men would be able to 'make a flying chariot' by which they could pass through the air.

As a tutor to the family of one of the Parliamentary leaders, Wilkins became acquainted with the rulers of the nation. He was active in scientific circles in London, which were in touch with the new rulers of the country and reflected their attitude to science. This attitude was that of merchants and landowners who looked upon their land less as a means of pursuing the feudal life than as a business for making profits. Many of these were more interested in mining and the exploitation of the minerals beneath their land than the cultivation of the land itself. They exhibited a keen interest in the invention and development of mining machinery, and especially of engines for raising water out of the workings.

The London scientists in touch with the Parliamentarians were centred around Gresham College, where they met for discussions. During the military operations of Parliament against Charles I, however, the College was commandeered for billeting troops. This made the meeting of scientists more difficult but did not damp their enthusiasm, stimulated by the great political events. The situation for scientists improved in 1647, when Cromwell appointed Wilkins

Warden of Wadham College, Oxford, with the aim of converting the University from a Royalist into a Parliamentary stronghold.

Wilkins attracted to Oxford many of the scientists who were finding conditions of work difficult in London. These included John Wallis the mathematician, and William Petty, an outstanding man of a new type, a founder of the science of statistics, who began to conceive the science which is required by modern trade and business. Besides the London men who gathered around Wilkins, and others, such as Robert Boyle, who settled in Oxford on his invitation, he had very gifted pupils, including Christopher Wren and Robert Hooke.

The discussions the scientists had had in London were extended in Oxford. Later on, when London became more settled, meetings were resumed at Gresham College. After exhibiting great scientific talents, Christopher Wren was appointed professor of astronomy at Gresham in 1657, when he was twenty-five. Isaac Newton in his *Principia* refers to him, together with Wallis and Huygens, as one of 'the greatest geometers of our times'. Newton made use of Wren's experiments demonstrating the laws of impact. Wren made a variety of interesting researches but did not pursue them very far, because he was soon drawn into the profession of architecture.

With the revival of Gresham College on the appointment of Wren and others, the scientists formed the habit of meeting after his lectures for further discussions. At one of these meetings in 1660, with Wilkins in the chair, the scientists proposed to form themselves into a society. When the approval of Charles II had been obtained the society was duly formed as The Royal Society of London. Wilkins, who as Cromwell's brother-in-law could not very well be president, was its first Secretary. Under Wilkins's influence especially, the Royal Society started a planned development of science specifically on the lines proposed by Bacon. Wilkins's pupil Robert Hooke was engaged by the Society to pursue experimental researches on topics suggested to him.

Robert Hooke was born in 1635, the son of a poor curate. He seems to have been a distant relative of Wren. He was a delicate boy, with a weak physique, which was a cause of his uncertain temper throughout his life. He showed remarkable talent from his infancy. He had an extraordinary memory, mechanical aptitude and talent for drawing.

Robert Boyle engaged him as assistant in experiments. He made an improved air pump, which Boyle used in his famous experiments on the properties of air. Hooke experimented on an extraordinary range of subjects. He made many experiments on model flying machines. He became interested in astronomy, and this led him to the problems of the measurement of time and the construction of clocks for determining longitude at sea. He invented the clock spring. He improved the barometer, making it suitable for general use in observing the weather.

Hooke was appointed professor of geometry at Gresham College in 1665. In the same year he published his great work on microscopy, *Micrographia*. Among the many discoveries recorded in it is the biological cell, which he first recognized in vegetable tissue. His picture of the louse became particularly famous. His study of the silk fibre, and how the silkworm makes it, led him to propose the invention of synthetic silk, by forcing a glutinous material through a small hole. His investigation of the properties of very thin sheets of glass led him to discover the diffraction of light. He observed the coloured rings caused by it, subsequently known as 'Newton's rings'. Besides his experimental researches, Hooke speculated on the mechanics of the solar system. He suspected that the planets were made to move in their orbits by gravitational forces which varied inversely as the square of their distance from the sun.

The founders of the Royal Society, accompanied by many other talented men, worked to produce a compact body of scientists starting on a clear-sighted programme of scientific development for both philosophical and practical ends.

Their work prepared the way for the emergence of Isaac Newton, who was born on Christmas Day, 1642, near Grantham in Lincolnshire. He grew up and was educated during the Commonwealth, but unlike his immediate scientific predecessors he did not reach manhood under it. He was sent to Cambridge in 1661, and so he started his adult life after the Restoration. Newton was the son of a yeoman farmer. His father died young, and his mother married a well-to-do clergyman. From his youth he had an assured income for life of £80 a year, which in those days guaranteed him a living. He was sent to the local grammar school, where he ultimately became

18. Hooke's Louse

head boy. He was quiet, thoughtful and did not like rough games. He was fond of making mechanical toys and reading scientific books.

As he showed no aptitude for farming, he was sent to Trinity College, Cambridge, to qualify as a clergyman. He did not show any particular talent until he came under the instruction of Isaac Barrow. This remarkable mathematician, Greek scholar and theologian was an ardent Royalist and courageous fighter. On the initiative of Wilkins, he was appointed to the newly created Lucasian chair of mathematics at Cambridge in 1663, when he was thirty-three. In his researches he had solved particular problems that involved the methods of the calculus, and he had advanced the study of geometrical optics.

Under Barrow, Newton's mind took fire. In the year following Barrow's arrival, Newton was awarded a scholarship. This led to his adoption of the academic life, instead of becoming a clergyman. He began to read Descartes's treatise on algebraical geometry, in which Descartes had invented the use of algebra for solving geometrical problems. Like the invention of a better symbolism for numbers, or of the computer, it gave method a larger place in the solution of problems, and thus greatly facilitated the advance of science. Descartes had invented his algebraical geometry as a means of calculating the quantities in Galileo's graphs of the motion of bodies.

Newton had already made notes on the theory of the Copernican system in 1661. Since then he had become acquainted with two of the sets of ideas, Galileo's mechanics and Descartes's algebraical geometry, to give it greater precision. At the same time, in following Barrow, he became equally interested in experimental and theoretical optics. He read Kepler's *Optics*, which inspired him to make the first reflecting telescope, the parent of the 200-inch reflector at Mount Palomar.

Then, in the summer of 1665, he was forced to leave Cambridge owing to the bubonic plague. He went to his Lincolnshire home at Woolsthorpe. During the next two years he spent more time there than in Cambridge. His mind was packed with new knowledge and ideas, upon which he meditated and experimented undisturbed.

Within the two years he had conceived the theory of gravitation, invented the calculus, discovered the binomial theorem, the general method of expression of algebraical functions in infinite series, and made the great experimental discovery of the spectrum of light.

19. Newton's reflecting telescope

Referring to this period, Newton subsequently wrote: 'All this was in the two plague years of 1665 and 1666, for in those days I was in the prime of my age for invention, and minded mathematics and philosophy more than at any time since.'

In 1669 Barrow resigned his chair in favour of his wonderful pupil, as he wanted to devote himself more to theology, which then had much greater prestige. Newton was now comfortably off, according to the values of the period. He had only to deliver twenty-four lectures a year. His first course of lectures was on optics. The Royal Society heard that they contained novel material, and so wrote to him for information. He replied by sending a description, and

20. The 200-inch reflecting telescope at Mount Palomar

replica, of his reflecting telescope. He was astonished at the excitement it caused, for he regarded it only as a trifle. He suggested that he should send them a paper which would be really worth-while. It would contain 'the oddest, if not the most considerable detection, which hath hitherto been made in the operations of nature'. Such words from a young man who had not yet printed anything were indeed magisterial, but well founded, and very characteristic of Newton. The paper he submitted contained his discovery of the spectrum of light.

The proof that light consists of bundles of rays of different refrangibility, so that any beam can be analysed into precise individual components, is, in the opinion of Heisenberg, the starting-point of modern theoretical physics, for it enabled the phenomena of light to be submitted to precise mathematical description and analysis. Newton's first published paper raised him at once from obscurity to international status. It also contributed to the beginning of diffi- culties in personal relations with other scientists, which grew with the years.

Newton's paper, published in 1671, owed more to Robert Hooke's *Micrographia* than he enjoyed admitting. Hooke, seven years his senior, definitely felt Newton had taken more than he admitted. Newton shrank from the insinuation and intimated that he wished to resign from the Royal Society. He apparently withdrew more into theological and alchemical research.

In 1679 Hooke became Secretary of the Royal Society. As such, he had to secure interesting papers, and so wrote politely to Newton, asking if he had any scientific news. Newton wrote a sarcastic reply, and added a titbit at the end to 'sweeten my answer', as he afterwards told Halley. He discussed what would happen if a bullet were dropped from a great height, without resistance. He suggested that it would approach the centre of the earth in an increasingly close spiral. Hooke, Wren, Flamsteed and others discussed this, and Hooke pointed out that the bullet ought to revolve round the earth in an ellipse. Newton was abashed at being corrected by Hooke, of all people. He angrily looked into the mathematics of planetary orbits, and satisfied himself that if a planet moved round the sun in an ellipse, then it followed that the gravitational force which kept it in motion must vary inversely as the square of the distance between planet and sun. He kept this to himself.

Hooke, Wren and Halley were still discussing the problem five years later, not having found the solution. In 1684 Halley went to Cambridge to consult Newton, and was astonished to learn that he had solved the problem years ago. Halley now set out to persuade the sensitive genius to develop and write out his theory of gravitation at length. Newton was forty-two, and Halley a very intelligent and persuasive young man of twenty-eight. Halley not only persuaded Newton to write the *Principia Mathematica Philosophiae Naturalis* (the Mathematical Principles of Natural Philosophy) but actually paid for its publication. Newton even spoke to Halley of the *Principia*, this greatest of all scientific books, as 'your book'.

Newton put the material of the *Principia* into shape in a period of about eighteen months. It contains the equivalent of a quarter of a million words. The first part consists of a statement of the laws of

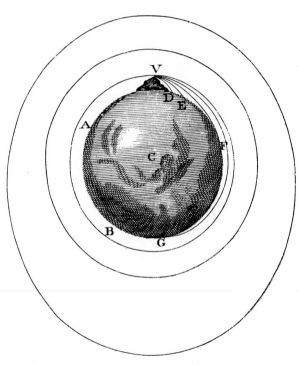

21. Newton's diagram of the path of an artificial satellite

motion, in which Galileo's work was extended and given fuller mathematical formulation. In the second part he analysed the motions of bodies in resisting media. This was necessary in order to discover whether the heavenly bodies were moving in a resisting medium or in empty space. In the cause of this he applied mathematics to the theory of gases and liquids. He showed that Boyle's law, according to which the volume of a gas varies inversely to its pressure, could be derived mathematically from an atomic theory of matter. He calculated the speed of sound waves, and checked his results with the echo which can be heard in one of the courts at Trinity College. His mathematical analysis of the motion of bodies in fluids led to the foundation of the science of hydrodynamics. He deduced the shape of a body which would give least resistance in passing through a fluid, and suggested that it might 'be of use in the building of ships'.

In the third part he applied his completed system of mechanics to the analysis of the movements of the heavenly bodies, conceived as masses of matter attracting each other according to the laws of gravitation. Newton worked out the theory of artificial satellites and in 1728, a year after his death, a diagram illustrating their orbits was published. The contrast of Newton's complete description of the physical universe down to the last detail, as then known, with the gropings of Copernicus, the divinations of Kepler and the faulty speculations of Descartes, was almost that of the superhuman with the human. Not the smallest thing in contradiction with its principles was discovered for two hundred years. Newton seemed to have lifted mankind into a new and higher region of knowledge. His universe appeared as a perfect mechanical watch, made and set going by the Creator, and then left to run by itself for ever.

Newton believed that the theological implications of his work were of the greatest importance. He considered that he had demonstrated that the world must have been made by a Rational Being, and that therefore God must exist. He never forgot, however, that his theory of the solar system gave the solution in principle to the most important practical scientific problems in the England of his day: the exact calculation of longitude, the theory of the tides and even of tidal levels in important British ports. Near the end of the *Principia*, he comments that his analysis 'abundantly serves to account for all the motions of the celestial bodies', and, he adds, 'of our sea'.

After the publication of the *Principia* Newton hungered for an official position. He was appointed Warden of the Mint in 1696 and Master in 1700, by his former pupil Charles Montague, later Lord Halifax. He carried out his duties with exemplary honesty and efficiency, though without any particular originality. He died in 1727, a rich man.

Newton did not publish his treatise on *Optics* until 1704, after Robert Hooke had died. The late publication of the work enabled him to include an appendix of scientific speculations, which he called *Queries*, that had occurred to him during his life, and which seemed to contain important truths he had not been able, or had not had the time, to demonstrate. He expressed ideas which foreshadowed thermodynamics and quantum theory. He speculated that atoms were combined into bodies by electrical forces, and that the nervous and muscular system operated by electrical messages. He guessed that the average density of the earth is about five and a half times that of water, which is almost exactly correct.

Newton's work completed the scientific development arising from the age of exploration and merchant endeavour. He had given the complete answer to the navigator's universe, and science paused in its speed of development for about a hundred years before it received a new impetus, comparable in power to that which had carried Newton to the summit of achievement.

11. New sources of power

The change of England from an agricultural into a trading country received a great impetus from Henry VIII, through his dissolution of the monasteries. These owned about a quarter of the cultivated land. Henry asserted that it was being wastefully managed, and gave it to energetic followers who could be depended upon to exploit it in a more profitable way. These new landlords became the parents of several of the statesmen who served Elizabeth I and provided her reign with such creative energy. Ambitious men now looked more to trade and finance as the means to power. Even the families who had owned estates for hundreds of years viewed them less as a source of food and clothing for themselves and their dependants, and more as profitable businesses supplying food and raw materials to the growing urban and industrial population. Successful traders invested their fortunes in land and copied feudal manners, but did not lose their original commercial attitude to possessions.

Thus the more far-seeing of the old aristocracy and the new landed magnates of commercial origin both engaged in the commercial and technical development of their estates. Some of the earliest and biggest of the projects which arose out of the cultivation of the land on a more business-like basis were drainage schemes, for making waste marsh productive. In 1630 the fourth Earl of Bedford formed a company for draining 95,000 acres of the Fens. They engaged the Dutch engineer Vermuyden to carry out the scheme. He cut channels around the higher pieces of ground, so that rain was drained directly off it and conducted to rivers, in this way preventing it from running into the marshes, which had previously acted like a large permanent sump. As a result, areas of marsh became dry and could be cultivated. His plans took twenty years to carry out. Since that time the acreage of drained marshland in the Fens has been increased to

700,000; providing much of the most fertile land in England, which usually produces double the normal crop.

These drainage schemes stimulated interest in the problems of surveying, excavation, hydraulic engineering and the development of pumps. These could be driven by windmills, whose irregularity of operation was not a fatal handicap in raising drainage water, which did not have to be pumped out at any particular time as long as it was raised. Landlords with mineral resources attended to their exploitation in the same more business-like spirit. Consequently in the middle of the seventeenth century business-minded squires who had found valuable minerals beneath their lands became keenly interested in machinery, and especially pumping machinery. A new source of power for driving pumps was required, stronger and more reliable than windmills.

The Marquis of Worcester was an outstanding example of these mechanically minded landlords. He published a book with the title *A Century of Inventions*, containing descriptions of a hundred mechanical devices, and in 1663 obtained a patent for raising water by steam. Like others he had thought of steam pressure as a new source of power. The problem was to make a device that could utilize it. He designed a pump by which water could be blown up a pipe with steam from a boiler. It operated in three motions, controlled by three cocks or taps, one in the steam pipe from the boiler, a second at the top of the delivery pipe and a third controlling the inflow of water to the bottom of the delivery pipe. By appropriate operation of the cocks, the steam blew the water up the delivery pipe to the higher level, thus raising the water. The Marquis's descriptions were rather vague; perhaps because he had not made all of his devices work, or because he was concealing the crucial details to frustrate imitators.

Steam pumps based on this principle were successfully introduced by Savery in 1698, for raising water for domestic supply in houses. They were not suitable for industrial use because they were inefficient and liable to breakdowns. The steam was in contact with the water, and condensed very quickly, with consequent loss in pressure. Attempts to make up for this by increasing the pressure led to explosions. A method of keeping the steam out of direct contact with the water was required. About 1690 the French inventor Denis

Papin, the inventor of the pressure cooker, showed how a piston could be raised in a cylinder containing a little water by applying heat

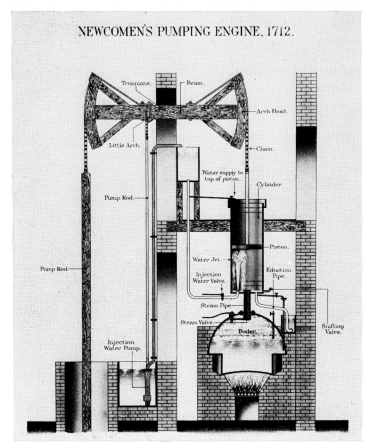

NEWCOMEN'S PUMPING ENGINE, 1712.

22. Newcomen's engine

to the outside of the cylinder. The water was converted into steam, which pushed the piston up.

The first effective industrial engine utilizing steam power was invented by the Devonshire hardware merchant Newcomen about 1702, who dealt in picks and shovels and other miners' tools, and was

H

directly aware of the pressing needs of the mining industry, in the Midlands as well as Devon and Cornwall. He successfully introduced Papin's piston into the mechanism of the steam pump conceived by Worcester and Savery. This made it efficient and strong enough to be a practicable industrial machine. It consisted essentially of a working cylinder containing a piston. The piston was lifted by steam from a boiler. When it had risen inside the cylinder the steam was cut off, and cold water sprayed into the steam. This caused the steam to condense, and the pressure of the atmosphere on the top of the piston forced it down, as in Guericke's experiment with the Magdeburg spheres (Fig. 10). The piston was attached to a lever or beam, so that as it went down, the other end of the beam pulled up a rod operating a pump at the bottom of the mine.

Newcomen's mechanism was the same in principle as the ordinary hand-pump for raising water from a well in the yard. He sought government support for the development of his engine. It seems that he was seen by Isaac Newton, who was convinced that, at any rate, Newcomen had an incorrect idea of how his engine worked. Nevertheless, Newcomen's engine worked. It introduced steam power to industry, in particular for pumping water out of metal mines and coalmines, and in spite of its low efficiency, it held the field for fifty years. This was because it could be run on waste coal, which at the pithead cost virtually nothing.

The Newcomen engine gave a big impulse to the development of metal mining in Cornwall, and coalmining in the Midlands, the North-East and Scotland. Attempts were made to apply it to driving mills, and even ships, but it was not convenient or efficient enough for these purposes.

Meanwhile, huge fortunes were being made by the new landlords. One of the most spectacular was Sir Hugh Smithson, the son of a Yorkshire squire with a business outlook. In 1740 he married Elizabeth Percy, the heiress of the Northumberland estates of the Percy family. He sank many pits on the family estates, whose rents rose from £8,607 in 1749 to £50,000 in 1778. The coal fuel for the rapidly growing domestic and industrial population of London was imported from Newcastle. Smithson, who became the first Duke of Northumberland, had a more splendid equipage than George III, which provided a striking illustration of the place and influence of the new industrial magnates.

Not even Newcomen engines could keep up with the voracious demands of the new industrial magnates, in whose mines coal-tubs were dragged from the coalface to the pit bottoms. The coal was carried up the mine shafts by hand, or by manually- or horse-operated cranes, which were insufficient. Colliery winding engines for hauling up the coal became necessary. This provided a demand for engines which could revolve wheels.

Comparable developments occurred in other centres with natural advantages, such as the Clyde–Forth district in Scotland. This contained coal deposits and convenient seaports, such as Greenock and Glasgow on the Clyde, and Leith on the Forth. Glasgow had a growing trade with the West Indies and America in sugar and tobacco, and Leith with the Baltic countries in timber and corn. Such was the size of the Glasgow trade that in the 1740s one Glasgow merchant imported one-twelfth of all the tobacco consumed by Europe.

These enormously rich Glasgow merchants had a club to which they invited the professor of moral philosophy in the University of Glasgow. This was Adam Smith, to whom they explained their principles of business. Smith worked these out and wrote them down in *The Wealth of Nations*, which became the textbook of the new world of business for the next hundred years.

Ports such as Glasgow became centres of population which prospered on the new trade. This stimulated the market for consumer goods and luxuries, such as textiles and whisky. The manufacture of these products posed problems in dyeing and distillation. Factories were built for converting imports, such as raw sugar and hides, into materials for food and clothing. Glasgow had the biggest tannery in Europe, and engineering shops were founded to make boilers for sugar refining. These industrial developments required a knowledge of chemistry and physics. The Glasgow manufacturers demanded that the University should start courses in chemistry which would fit their sons for the management of their factories. The eminent professor of medicine, William Cullen, started courses in chemistry and built a chemical laboratory for experimental work, specifically to meet this demand. Glasgow University still possesses the accounts of 1747 and 1748, in which Cullen was sanctioned to purchase chemical books and materials for these purposes.

As M. Daumas has observed, in the seventeenth and early eighteenth centuries chemists and their work were generally despised.

> 'Chemists had to stoke furnaces, they worked with evil-smelling substances, their clothes were generally covered with burns and stains, and their experiments were the source of many public complaints. All this was gradually changed as the study of chemistry began to offer increasing financial rewards, and as laboratories became better equipped.'

Cullen was one of the greatest physicians of the period, and was primarily interested in chemistry from the medical point of view. But he met the new industrial demand by investigating the chemistry of bleaching and the purification of common salt. The process of distillation was fundamental in the growing industries, especially in the manufacture of whisky. Distillation depends on evaporation, and Cullen's attention was attracted to this phenomenon. While reading an account of recent chemical and physical experiments, he was led to suspect that when water and other fluids evaporate they produce a fall in temperature. He directed one of his pupils to dip a thermometer rapidly in and out of a liquid, and increase the rate of evaporation by moving it about in the air very quickly. By these methods, he succeeded in producing a drop of 44 degrees in temperature with alcohol. He then made experiments on increasing the rate of evaporation, and hence the degree of cold, by placing water under an air pump and reducing the pressure over it. He succeeded in producing ice in this way, and became the inventor of the first refrigerating machine. This is a form of heat engine; thus, from the beginning of industrial and scientific development at Glasgow, research had been impelled in the direction of heat engines.

Among Cullen's pupils at Glasgow was Joseph Black, the son of John Black, a Scotch-Irish wine importer from Belfast, who had settled in Bordeaux. John Black sent his son Joseph to Edinburgh to study medicine under Cullen, who had moved to that university, but Joseph found himself more interested in Cullen's chemical lectures. Cullen recognized his exceptional gifts and refused to regard him as a student, treating him as his personal assistant.

Cullen's chemical course had been directly inspired by the needs of the new Glasgow chemical industries. Hitherto, chemistry had been very much influenced by the needs of medicine. These were qualitative rather than quantitative; doctors were primarily concerned with curative effects rather than with the exact quantities of drugs used. The attitude in chemical industry was different. The amounts of raw material used were large, and so were those of the fuel consumed in the manufacturing processes. Consequently, costs of raw materials and fuel were substantial, and profits depended on their economic use. Thus the development of industrial chemistry inspired the exact measurement of materials involved in chemical processes and the amount of fuel consumed in providing the heat necessary to promote them.

Joseph Black absorbed these attitudes as a student, and had the ability to carry them into both chemistry and physics. Before he was thirty years old he had invented quantitative chemical analysis, and laid the foundation of the quantitative theory of heat by his discovery of the specific heat of substances, the amount of heat necessary to raise a unit mass through one degree of temperature, and latent heat, the heat required to effect a change of state, such as from liquid to vapour, without raising the temperature. This gifted man was appointed professor of medicine and lecturer in chemistry at Glasgow in 1756, when he was twenty-eight years old.

One of the wealthy Scots merchants, Alexander Macfarlane, who had maintained a fine astronomical observatory in Jamaica, bequeathed his instruments to Glasgow University; they arrived there in packing cases and were deposited in storerooms. An instrument-maker was required to unpack them and put them into working order. The professor of classics, Muirhead, had a young relative named James Watt, who was an instrument-maker and was having difficulty in making a living. The University was persuaded to engage the young man as instrument-maker to the University, and give him the task of putting the bequest of astronomical instruments in order.

12. The invention of the steam engine

James Watt was appointed instrument-maker to the University of Glasgow in 1757, when he was twenty-one. His origins were not obscure. He was descended from a family of Aberdonian practical mathematicians and teachers of navigation. He came from the main stream of scientific inspiration in the age of exploration and trade, which had led to the triumph of Newtonian science. He grew up in a family sitting-room with a picture of Newton hung on one wall and Napier on another. His grandfather had settled in Greenock to practise his profession in the port, growing with the West Indian trade. He was followed by his son James, the father of the engineer, who conducted business as a ships' chandler, supplying navigational instruments, and was the owner of a small ship.

James Watt was intended by his father to inherit a considerable business. He was therefore neither apprenticed to a craft nor sent to university. However, the family fortune was ruined by the loss of a ship at sea. Owing to his age, James Watt was no longer acceptable by the Glasgow guild which included instrument-makers, and so was sent to London to secure a training surreptitiously, without membership of a guild. When he returned to Glasgow in 1756 he was not allowed to open an instrument shop in the city. This regulation did not apply to the grounds of the University, which was specifically free from the jurisdiction of the guilds owing to an order made by the Pope in 1451. When Watt opened his instrument shop in the University in 1757 he was twenty-one; Joseph Black was twenty-nine and Adam Smith thirty-five. There were several other remarkable professors. His relative Muirhead was one of the editors of the great Foulis edition of Gibbon. Glasgow was then one of the creative intellectual centres of the world.

While Cullen and Black were starting the scientific education of the future managers of the new technical industry, their colleague John Anderson, professor of natural philosophy, was attending to the scientific education of the craftsmen required in the new industry. He threw his classes open to artisans, giving them permission to attend in their working clothes. He gave lectures on scientific and engineering principles, illustrated by experiments and working models. Anderson subsequently left his instruments, books and estate to found an institution for the technical instruction of working men. This was the Andersonian Institution, which later became the Royal Technical College and is now the University of Strathclyde. By virtue of his efforts, Anderson may be regarded as the founder of technical education in Britain.

Among the models he used at his lectures was one of a Newcomen engine which, however, did not run properly. It was given to Watt to see whether he could make anything of it. He tried various alterations until he had got the engine to run continuously. Watt subsequently said that, unlike a 'mere mechanician', he did not leave it at that; he set about trying to find why it would not work. He was twenty-seven years old when he started on this research. He had been instrument-maker to the University for six years, and his workshop had become a meeting place for creative scientists, who enjoyed discussing scientific questions and instruments with the ingenious and well-informed craftsman. The gifted Professor Black acquired the habit of dropping in and handling his instruments, whistling to himself as he made delicate adjustments.

In this atmosphere, Watt's genius acquired scientific habits. He discovered that the model would not work because of scale effects. It was an exact scale copy of a full-size Newcomen engine. In such a model the ratio of the area of the walls of the cylinder to the volume is very much greater than in the full-size engine. Consequently, the rate of loss of heat from the model cylinder was much greater than in the full-scale engine. The model boiler could not supply steam fast enough to make up for this effect, and so the engine stopped after a few strokes. Watt now started a systematic study of the movement of heat in each of the engine's operations. He found that the cylinder of the model was made of brass, which conducted the

In 1765, JAMES WATT,
In working to repair this Model,
belonging to the Natural Philosophy Class in the University of
Glasgow, made the discovery of a
separate Condenser, which has
identified his name with that of the
STEAM ENGINE

23. The model of a Newcomen engine repaired by James Watt

heat away more quickly than the cast-iron used in the full-size engine.

He then tried to trace what happened inside the cylinder of a New-comen engine, utilizing Cullen's discovery of the effect of evaporating water under reduced pressure. He tried to improve the vacuum caused by the condensation of the steam by the spray of cold water. He made the jet of cold water larger, but found that while this improved the vacuum, more steam was required to warm up the cylinder in the next stroke. His measurements showed that it would be a very great advantage if the condensation of the steam could somehow be done without cooling the cylinder, but in spite of intense efforts he could not at first see any way of doing this.

He investigated the effect of temperature and pressure on the boiling point of water and plotted his results on a curve, in order to find the best temperature and pressure conditions for the running of the engine. He found that when a given volume of water was turned into steam it occupied a space 1,800 times greater. This enabled him to calculate the volume of steam consumed in each stroke of the engine, and he was astonished to find that it was several times that of the cylinder. He discovered, too, that a remarkably small quantity of steam could raise water to its boiling point; in fact, it could raise six times its own weight of cold water up to boiling point. He told Black of this discovery, who explained to him that it was an example of the transference of latent heat in a change of state. As a result of these researches, Watt had an entirely new insight of a quantitative character into the operation of a Newcomen engine. It gave him an exact and concrete grasp of the low efficiency of the engine, and of the great gain in economy that was to be won by conducting the condensation without alternate heating and cooling of the cylinder.

He thought about this problem for two years before the solution flashed into his mind while walking across Glasgow Green on a Sunday morning. He suddenly saw that it would be possible to have a separate vacuum chamber into which the steam could be exhausted and condensed. Within a few hours he had constructed in his imagination ways of doing this. He saw that it would not be possible to prevent steam from leaking round the piston by covering it with water, as in the Newcomen engine, for the cylinder would always be hot. This led him to let steam into the cylinder above the piston and use its pressure to drive the piston down, instead of using the pressure of the atmosphere.

Thus he invented the steam engine proper, for the Newcomen engine used steam only indirectly. He was aware from his previous steam measurements that his engine would be about four times as efficient as the Newcomen engine. Within two weeks he had made a working model of his engine, which is now in the Science Museum in London. James Watt's invention of the steam engine proper is the most important of modern times. As much as Copernicus and Newton, and perhaps even more, his work marks the frontier between ancient and modern history, because it opened the way to the production of unlimited power. The limitations of manual and animal, wind and water power, and even of the Newcomen atmospherically driven engine, prevented any revolutionary expansion of human production and endeavour.

Watt's genius in developing his engine and the engineering required for the purpose were not less striking than the invention itself. Before his time engineering had been a craftsman's job. Newcomen engines were built by craftsmen from materials assembled on the site, in the same manner as most domestic houses are still built today. These men worked by rule-of-thumb, to the nearest eighth of an inch. It was one of the merits of the Newcomen engine that it would work, though so roughly made. When Watt tried to build an engine of an industrial size, with the steam pressure acting directly on the piston, he found that contemporary mechanical engineering was unable to make an engine with sufficient precision to take adequate advantage of his separate condenser invention. He had to embark on a long and difficult task of expensive engineering development; and to obtain the financial backing that would enable him to persevere with the problem.

His first business supporter was Dr John Roebuck, the inventor of the lead-chamber process for manufacturing sulphuric acid. This had spectacularly reduced the price of the most important of industrial chemicals. Roebuck had seen great industrial possibilities in the River Carron area of the Firth of Forth. He founded an industrial complex there, in which coal was to be mined and iron ore smelted, and a range of products manufactured, from cannon to boilers. It was to be operated on the most advanced contemporary scientific and technical principles. Roebuck ran into serious flooding difficulties in his mines, and became urgently concerned with the problem of mine pumping. He wanted something more powerful than New-comen engines. As a scientist, he fully grasped the magnitude and significance of Watt's invention of the separate condenser, and so enthusiastically encouraged and financed him. But Roebuck presently got into financial difficulties, and Watt had to find other support.

As in other industrial centres, the more progressive manufacturers in Birmingham were looking for increased sources of power. Their outstanding leader, Matthew Boulton, who was manufacturing metal goods on rationally organized lines, was looking for an engine which would keep his works running steadily, so that he would have the advantages of regular production. He used water power, which was liable to fail in dry seasons and interrupt production. His idea was to have a pumping engine which would pump the same water over and over his water-wheel when the flow from the stream had stopped. Boulton's model works and progressive technical personality attracted men of talent. Benjamin Franklin became one of his friends and discussed technical problems with him. The leading physician in the Midlands, Erasmus Darwin, helped in the same way. Franklin recommended Boulton to patronize Dr William Small, a Scottish doctor and physicist, who had been a professor in Virginia, where he had taught Thomas Jefferson and 'fixed the destiny of my life', as Jefferson subsequently said. He had to leave Virginia because of ill-health and so was glad to settle in Birmingham under Boulton's wing. Small knew his fellow-Scot James Watt, and, through Small, Watt came to visit Birmingham. On the first occasion Boulton was away, and Watt was received in particular by Erasmus Darwin, who at once recognized his genius and diagnosed his temperament.

On the next occasion Watt met Boulton, and these two outstanding men at once perceived that they had complementary personalities; Watt had the genius and Boulton the business sense. Boulton conceived the plan of securing the patent for the new source of power in all countries and then drawing royalties on it from all the world. Boulton founded a separate firm, Boulton and Watt, to manufacture the steam engine. It became the most famous engineering firm of its day. In it, modern engineering-drawing of machinery, the planning of the layout of machines in workshops, work-study and industrial insurance were promoted and to a considerable degree created. The Boulton and Watt works acquired a talented staff. This included William Murdock, who lit the works by coal-gas. Equally gifted was the engineer James Southern, who, with Watt, invented the indicator diagram. This crucial invention draws a graph of the pressure and temperature changes which occur in the cylinder of a steam engine during the stroke of the piston. By it, the engine is made to record automatically the physical changes of the steam going on in itself. The young French physicist Sadi Carnot showed that the cycle of operations in the steam engine enabled the efficiency of a perfect engine, working within a given range of temperature, to be calculated exactly.

Watt devised an absolute measure of the power supplied by an engine. This was necessary for commercial reasons, in order to measure the engine's commercial value and hence the price to be charged for it. He defined the horse-power for this purpose, as the power required to raise 33,000 lb through one foot in one minute. He invented locked meters which could be attached to his engines, and automatically register the amount of work they had done. Watt's precise measurement of the amount of work which his engines did led to the concrete scientific conception of energy, and to Joule's measurement of the mechanical equivalent of heat, and hence to the establishment of the principle of the conservation of energy. The combination of the principle of the conservation of energy with Carnot's cycle led to the foundation of the science of thermo-dynamics.

Thus the modern conception of energy, and the science for dealing with it, was inspired by Watt's steam engine. Even this is far from all that has flowed from Watt's work. He developed the principle of the governor to regulate the speed of his engines. This contained the first important application of 'feed-back', by which a mechanism is made to control itself. The first major advance of the theory of 'feed-back', on which cybernetics, or the science of self-controlling machinery and automation, depends, was made by James Clerk Maxwell, in his mathematical analysis of the operation of Watt's governor.

13. History speeds up: evolution

The development of the steam engine was not only a supreme triumph of the use of science for advancing industry; it also undermined the ancient static view of history. The unlimited development of power was an entirely new possibility. It introduced a cause of change which would be continually increasing in size. History could begin to take on a rapidly moving dynamic aspect. The Industrial Revolution and the steam engine showed that fundamental changes could take place in the customary order of things. This prepared scientists to recognize such changes in the structure of the earth, in plant and animal life, and in the whole of nature. It became possible to discover the Theory of Evolution.

Up to Watt's time human production and population had increased at so slow a rate that life and the world seemed essentially static. Isaac Newton, the supreme scientist of the preceding age, had regarded the universe as a clock-like mechanism created by the Almighty some four thousand years ago and thence running on of its own accord. He spent a great deal of his time and ingenuity in trying to fit the events of history into this brief four thousand years during which he supposed the universe to have existed.

The impulses which led to the supersession of this static outlook came from the centres of the new development of industry and power. James Hutton, a geological friend of Watt, started the revolution in geology by providing good evidence for the belief that geological forces, similar in character to existing forces, had acted over very long periods of time, and he explained the changes in the surface of the earth as due to internal heat. He conceived the earth as a heat engine which went through a series of motions protracted over immense ages of time. His ideas were supported and substantiated by the geologist Lyell.

Watt's friend in Birmingham, Erasmus Darwin, at the heart of the steam-power development, propounded a theory of the evolution of the whole of nature, including plant and animal life. He was one of the chief founders of a mode of thought which was recast, developed and demonstrated with revolutionary power by his grandson Charles Darwin. Erasmus Darwin was born in Nottinghamshire in 1731, and sent to Cambridge to study medicine. There he developed the habits of an English gentleman and made little progress in medicine, and so was sent to Edinburgh to continue his medical studies. He arrived in 1754, just at the time when Joseph Black had invented quantitative chemical analysis, in the course of his investigation into the properties of alkalis. Edinburgh was at the height of its intellectual activity, and formed Erasmus Darwin's scientific outlook.

He set up as a doctor in the Midlands, where he soon acquired a big practice among the country gentlemen and the new industrial magnates such as Wedgwood and Boulton, who were glad to avail themselves of his ideas and his scientific and technical judgement, besides his treatment for their illnesses.

He sent to Boulton in 1765 a design of a steam carriage driven by two cylinders. It was too advanced and was not taken up, but he had the technical background to appreciate Watt's invention when he first met him two years later. For Wedgwood, Darwin invented a horizontal windmill to grind colours. He helped in the design of canals, promoted by Wedgwood for the transport of the increasing volume of industrial products. In the course of this, he invented the double-lift for raising barges over hills, a device adopted extensively in Germany after 1930. Erasmus Darwin's son Robert married Wedgwood's daughter Susannah, who became the mother of Charles Darwin.

Among other engineering sketches left by Erasmus were continuous-flow rotary pumps, and water and steam turbines. He designed an advanced form of water-closet. He designed a speaking-machine which could pronounce simple words. Particularly striking was his application of the centrifuge to medicine. He suspected that the condition of lunatics might be relieved by reducing the pressure of the blood in their heads; he designed a large centrifuge for whirling the patient at the end of a long arm, which caused blood to flow from his head. James Watt made the engineering drawing for this device.

Such centrifuges are now part of the apparatus used for training astronauts to withstand the variations of gravity in rockets and artificial satellites.

Erasmus Darwin was particularly interested in meteorology and the physics of cloud formation of the atmosphere. He gave the first adequate account of adiabatic expansion and compression, in explaining how clouds are formed. He noted the existence of what are now called warm and cold fronts, and he proposed to measure the north–south drift of the air by means of an air meter, consisting of a horizontal cylinder pointing north–south and containing a recording vane. He was the first to form correct ideas on the structure of the atmosphere; he suggested that the outermost parts consisted mainly of hydrogen. He held that the aurorae were electrical phenomena occurring at heights greater than thirty-five miles.

He organized his extraordinary array of knowledge especially in two long scientific poems, entitled *The Botanic Garden* and *The Temple of Nature*. In the former he gave a summary of contemporary science, in verse and in prose footnotes, explaining the contributions of Watt, Priestley and Hutton, and their significance. In *The Temple of Nature* he sketched a theory of the evolution of man and human society from microscopic specks which first took shape in the primeval seas. The modern idea of the origin and evolution of life resembles his theory.

The great writers of Erasmus Darwin's time were well aware of his contribution. Coleridge described him as 'the first *literary* character in Europe'. Like Wordsworth and Shelley, he was deeply indebted to him for many ideas. At the beginning of the Industrial Revolution there was no sharp division between science and literature. This developed as the new industrial social order became more complex and the application of the division of labour extended. The different activities in life tended to become more sharply defined and the differences between these activities hardened. Each activity tended to be pursued more as an end in itself. Writers became 'literary', and considered science and business outside their field. Businessmen became exclusively concerned with profits, and scientists considered that literature was outside their field. Soon after Erasmus Darwin died in 1802 these views became conventional and established. Men who matured in the early years of the nineteenth century, including Erasmus's own grandson Charles, took them for granted. To the new generation, Erasmus Darwin seemed a fantastic amateur; they felt that all his work required redoing, on what to them was a proper professional basis.

Charles Darwin was born in 1809. He inherited the qualities of the Wedgwood rather than the Darwin side of his family. Like Josiah Wedgwood, he was very persevering and systematic in research, and an excellent businessman. Charles Darwin left a fortune of £287,000, whereas his grandfather, who had surprised Charles by leaving comparatively little, had received large fees from his rich patients but treated numbers of poor people without charge. Charles was sent by his father to Edinburgh to study medicine. He was upset by the unpleasantness of the medicine of those days and made poor progress in medical studies. His father took him away and sent him to Cambridge to read for the Church. Charles was equally unsuccessful in these studies, but he acquired an extraordinary talent for collecting beetles. He kept rare specimens in his mouth until he had an opportunity to preserve them. His collecting skill attracted attention, and he was invited to go on excursions with the leading naturalists in the university.

After taking a modest degree he read at leisure. Among his books was Humboldt's narrative of his travels in Central America. This suddenly fired his imagination. Together with this, he read John Herschel's *Study of Natural Philosophy*, which gave him a clear grasp of scientific method. These books opened his mind, and he saw that perhaps he might become a scientist, and escape medicine and the Church. Shortly afterwards one of his Cambridge teachers told him that Captain Fitzroy was organizing a voyage round the world and desired to have a naturalist to accompany him. Would he be prepared to go? Charles was overwhelmed, and consulted his father, who was against it. Then he consulted Josiah Wedgwood, his uncle and son of the great potter, who strongly advised him to go.

Charles, then twenty-one, went to see Fitzroy, who was only twenty-five. He was an illegitimate descendant of Charles II and a nephew of Castlereagh, like whom he ultimately committed suicide. Fitzroy was a fine seaman, of violent but forthright character; he was

I

a fanatical believer in Christianity and slavery. The object of his voyage was to survey the coasts of South America on behalf of the British government, and he returned with a splendid collection of original charts of coastlines and harbours. Fitzroy's ship, the *Beagle*, displaced only 235 tons and had a crew of no less than seventy; accommodation was extremely tight. He set out at the end of 1831. Charles succeeded in sharing a cabin with this extraordinary man for several years; he had great patience and self-command.

Darwin was thrilled by his first sights of tropical plants and animals. They far surpassed anything he had imagined. He kept a careful diary, which from the first showed his deep preoccupation with the scientific problems and significance of what he saw. His genius was natural, but its remarkable discipline seems to owe much to his early reading of John Herschel; he knew from the beginning how to manage his mind and his material. He had with him volumes of Lyell's treatise on geology, which was in process of publication. This stimulated his reflections on the astonishing sights of the Andes. He experienced an earthquake while he was there and observed its tremendous effects. He pondered on the forces which had caused it and which must have caused similar effects in the past. He was profoundly excited by the presence of volcanoes in Tierra del Fuego, and the extraordinary difference between the primitive savages and European man. He wondered whether they had been like that since the creation of the world, or whether civilized man was the result of a long improvement on that original savage state. Darwin was also deeply impressed by the huge quantities of fossils of extinct animals.

At length, after more than three years of wonders, which seemed all the more startling against the background of his memory of the quiet English scene, the *Beagle* approached the Galapagos islands, a group on the Equator about 800 miles west of Ecuador. The governor of the islands remarked to him that the tortoises on the various islands were different, and one could tell from its appearance from which island any tortoise had come. Darwin now found that this applied to the birds as well. He thought about the significance of these observations and soon formed the conception that these varieties of animals were the descendants of fewer kinds, which had found their way to the different islands and then propagated. The isolation in which they found themselves caused their descendants to acquire somewhat different characteristics according to each particular island population. This was the most pregnant of the wide range and large number of suggestive observations that he made, and had the greatest part in stimulating him to conceive his Theory of Evolution.

After arriving back in England he started a new notebook in 1837, with the title *Origin of Species*. He put down in it reflections on the material of his great voyage and other facts which seemed to have a bearing on the problem. The fact of the evolution of living organisms from simple into more complex types seemed clear, but he could not at first imagine any mechanism by which it could be brought about. In 1838 he read Malthus's *Essay on the Principle of Population*, in which it was argued that a population tends to increase in geometrical ratio, while the means of subsistence increase only in arithmetical ratio. Thus the difficulty of providing food was a constant check on the growth of population. This suggested to Darwin that under such conditions only organisms with favourable variations would survive, while those with unfavourable ones would be destroyed. This mechanism, subsequently described as the principle of natural selection, provided him with the solution that he was seeking.

Darwin now planned an immense work which was to give a complete and detailed demonstration of the theory of evolution by means of natural selection. In 1858, when he had already been at work on it for twenty-one years, he heard that the naturalist Alfred Russell Wallace had arrived at a similar conception on the basis of his observations in the Malay Archipelago. Most fortunately Darwin and Wallace recognized the independence of each other's work. They published a short joint paper later in 1858, stating the essence of their views.

Darwin's friends urged him to publish a summary of the work that he had been preparing during the last twenty-one years. He did this rapidly, and published it in 1859 under the title: *On the Origin of Species by means of Natural Selection or the Preservation of Favoured Races in the Struggle for Life*. This famous work, which occupies a place in the history of science comparable with that of Newton's *Principia*, was merely an exposition, in non-technical language which any educated person could read, of the significance of the huge mass of observations and thoughts which he had accumulated during the previous quarter of a century.

As with Newton, Darwin's masterpiece was not his only great work. He wrote a series of volumes in which he applied the new theory to different aspects of organic nature. In his *Descent of Man* he applied it to the evolution of man, and virtually founded the modern science of anthropology. In his *Expressions of the Emotions in Man and Animals*, he did the same for the science of psychology. In his *Variation of Animals and Plants under Domestication*, he began to place the science of inheritance, or genetics, on a scientific basis. To show that he was no mere speculator, as some said his talented grandfather and other notable men had been, he published huge technical monographs on barnacles and coral reefs and the fertilization of plants.

After this tremendous exhibition of thought and observation there could no longer be rational doubt of the fact and operation of the principle of evolution. It was no accident that this was the achievement of a descendant of those men who led the technical and scientific developments of the Industrial Revolution.

14. The search for materials and the scientific study of the earth

During the Middle Ages, when European society was largely organized in self-contained communities, the things that were obtained from outside were rarities, such as gold and the spices needed to make poorly preserved food more palatable. These were small in bulk and high in value, and offered big profits to bold travellers, who in search of them discovered the routes across Asia, around Africa and to America. These early voyagers behaved like bandits to technically backward peoples, taking their gold by force if they could do so with impunity. With the growth of European population and trade in the seventeenth and eighteenth centuries, food and raw materials became greater sources of wealth. There was more profit to be made out of sugar, tobacco and cotton for the many than gold and jewels for the few. This inspired a more systematic ransacking of the earth, to discover new commodities, and minerals, plants and animals which might provide fresh resources for the growing population and manufactures.

In Britain the supremacy of the commercial outlook which followed the Parliamentary rising was reflected in the foundation of the first scientific institution by a British government. This was the Royal Observatory, established at Greenwich in 1675. It was specifically for the pursuit of astronomy as a means to the improvement of navigation.

Talented sons of City merchants took up the study of the science. Foremost among these was Edmund Halley, who was born in 1656, during Cromwell's rule. His father was a rich soap-boiler and his grandfather the owner of many public-houses. Halley experimented with magnetism when a boy, and discovered for himself that the direction of the earth's magnetic field in London was subject to change. His father bought him astronomical instruments, and he studied geometry and astronomy. Before he was twenty he perfected the work of Copernicus and Kepler by producing a conclusive proof that the planets move in an ellipse with the sun at one focus.

Up to this time astronomy had been based almost entirely on observations made in the northern hemisphere. It was evident that the southern skies should be mapped equally well. Halley proposed that he should carry out such a survey. His father enthusiastically supported the idea. He settled a handsome income on his son, and secured the support of the government and the East India Company, who were interested in safe navigation. The Company gave Halley a free berth for a voyage to St Helena, and he sailed, just after his twentieth birthday, to observe the southern skies from that isolated and distant island.

The young astronomer recorded the positions of 341 stars. They formed the first catalogue of stars made with telescopic sights. Among other observations, he recorded the first complete transit of the sun's disc by the planet Mercury. This led him to point out that observations of transits of Venus would provide the most accurate method then known of calculating the distance of the sun from the earth, which is one of the fundamental units of astronomy.

After this he spent two years travelling in Europe and conferring with the leading astronomers. Besides completing the observation of the heavens, he aimed at doing the same for terrestrial magnetism, so that it could be described accurately and in detail for the benefit of the world mariners. In doing this, he sketched a theory of the origin of the earth's magnetism, which is similar in character to the one accepted today, and he invented symbolism for handling masses of statistical facts, which is also still in use. Through these researches he was led to found the study of the physics of the earth as a whole, or geophysics, now pursued on a world scale by such organizations as the International Geophysical Year.

As the first of Newton's disciples, he applied the new theory of gravitation to the calculation of the paths of comets. He forecast that the impressive comet of 1682 would reappear about 1758. This became known as Halley's Comet, and provided the first major proof of the theory of gravitation by forecast.

The computing which Halley's calculations required caused him to invent improved mathematical methods for dealing with statistics. He applied these methods to vital statistics of births and deaths, in order to disprove the relation between the stars and human life, and thus undermine the influence of astrology. As a result of this research, he founded the mathematical theory of life insurance.

In 1698 he was sent by the British government on a new expedition to survey the direction of the magnetic compass over the Atlantic Ocean, for the benefit of navigation. He was not a professional seaman, but he navigated his ship to the frontiers of Antarctica, where he came upon great islands of ice, and succeeded in returning home safely, with immense collections of data for charting the world's magnetism.

Halley made many other contributions. Lagrange learned from his works how to develop the modern method of applying mathematics to physical problems. Then, at the age of sixty-two, he demonstrated that certain of the 'fixed' stars must have moved since ancient times. This indicated that the stellar universe was changing its shape and going through some process of development. It was the beginning of modern cosmology.

Halley died in 1742 at the age of eighty-six. His comet duly appeared some sixteen years later, conferring wide fame on his memory. Scientists were mindful of his suggestion that the transit of Venus, forecast for about 1768, should be carefully observed, in order to measure the sun's distance. They approached the government for support for an expedition to Tahiti in the Pacific Ocean, for making the observations. The government gave a grant, and provided a ship and crew. They appointed as captain Mr James Cook, a very able seaman, who hailed from Whitby and was the son of a Yorkshire farm labourer.

Cook was not then a commissioned officer, and therefore not regarded as a gentleman. He had distinguished himself by making remarkably accurate charts of the River St Lawrence in the face of the enemy, to facilitate the attack by the British fleet which led to the capture of Quebec and the conquest of Canada. The contrast between Cook and the great explorers of the earlier period, such as Drake and Raleigh, was striking. They belonged to two different orders of society, with different aims, conceptions and methods. He was as staid and business-like as Drake and Raleigh were romantic and piratical. He was their equal in boldness but in a different way. He did not fight unless he had to, but he performed almost incredible feats of seamanship. He piloted his ship more than a thousand miles through the uncharted Great Barrier Reef, off the east coast of Australia, by continuous sounding with line and lead, threading his way through the coral reefs, never more than a few fathoms from wreck and destruction.

Cook first went to sea as a ship's boy in a Whitby coal-boat, and he chose one of these rough but seaworthy ships for his voyage, renaming it the *Endeavour*. Scientific staff accompanied him to make the astronomical observations of the transit of Venus from Tahiti, and the wealthy Lincolnshire landowner and naturalist Joseph Banks, then twenty-five years old, joined the party at his own expense, bringing with him nine assistants and a splendid collection of scientific equipment. They were to make systematic collections of plants, animals and minerals, and collect information on the peoples, in the various lands they visited.

The *Endeavour* reached Tahiti in April 1769, and the astronomers observed the transit of Venus. Meanwhile Banks and his assistants had been busily making naturalists' surveys and studying the peoples in the lands which they had touched. Cook sailed on to New Zealand, where Banks noted that it should be possible to cultivate European crops. From here Cook proceeded to the exploration of the Australian coasts. At one place Banks found so many new plants that he named it Botany Bay. Cook safely brought his ship home after two years, with his mission completely accomplished. The transit had been observed, and he had himself made scores of meticulous surveys of unknown coasts. Banks had returned with 800 new species of plants, and had perceived the possibility of colonizing New Zealand and Australia.

George III was but one of the many who were thrilled by the story of the voyage. He received Cook and Banks. He was himself a farmer and stockbreeder, and found himself at ease with Banks, whom he made President of the Royal Society in 1778. Banks remained in this position for forty-two years, guiding the British scientific world with a definite and fruitful policy consonant with the needs of the mercantile age, which had reached its climax and was beginning to be superseded by industrialism. Banks said that his voyage with Cook was the first specifically scientific voyage of discovery, and the forerunner of the scientific voyages now regularly organized to discover the contents and processes of the earth as a whole.

Through Banks's influence, George III founded Kew Gardens. This became the centre of information and of the exchange of plants for the British empire. It was due to him that the tea plant was introduced from China into India and Ceylon. He had Captain Bligh sent on the famous voyage of the *Bounty*, the aim of which was to introduce the cultivation of bread-fruit trees from Tahiti into the West Indies. Banks's utilization of science in empire-building impressed Napoleon, who was willing to listen to intercessions from him that scientists from either side should not be molested by the combatants in the war between the English and the French.

As President of the Royal Society, and through his personal authority, Banks became the adviser to the state on science. He nominated persons for government scientific committees.

Isaac Newton and Joseph Banks were the greatest presidents of the Royal Society in the mercantile period; Newton was supreme in adapting astronomy and mathematics, and Banks natural history and descriptive biology, to the needs of the age.

The increasingly pressing requirements of industrialism in the latter part of Banks's presidency, at the beginning of the nineteenth century, called for a new policy for science. This came from men of the next period, whose inspiration arose primarily from the industry with which they were personally in contact, and less from the motives of overseas trade and exploration. Mercantile men were interested in materials, whether finished goods such as calico from India or raw products such as timber from Russia, as a medium for trade. They searched the world for things in which they could trade. The industrialists were more concerned with the properties of materials, and the processes by which they could be changed into desired products. Their interest was therefore in the properties of matter, and how it could be changed; that is, in physics and chemistry, rather than in natural history and exploration by which the materials of trade might be found ready-made in some distant part of the earth.

At Glasgow scientists were specifically asked to attend to the kind of scientific information which industrialists required. They began to conceive chemistry and physics in terms of industrial ideas. In their manufacturing processes they were particularly concerned with the continuous properties of materials. They thought of liquids and gases as continuous fluids, and solids as congealed continuous fluids. From the beginning of the Industrial Revolution, about 1750, chemists and physicists became less interested for a time in atomic theories of matter. These theories were not yet sufficiently developed to throw much light on chemical processes. They were not revived until there were enough chemical and physical facts to provide an adequate basis for them.

Joseph Priestley (1733–1804) was outstanding in discovering new chemical facts of a qualitative character. He was the son of a York-shire cloth-dresser and weaver, and took the attitudes of such a craftsman into experimental chemistry. He worked in his own house, performing his experiments in the kitchen and heating his apparatus on the kitchen fire. He developed the method of handling gases in vessels inverted over troughs of water. Then Priestley pursued experiments in the kitchen garden, and grew sprigs of mint in bottles. This led him to the great discovery that at night plants have the power of restoring to used air its capacity to support life. Before he began his researches, chemists clearly recognized only three gases, air, carbon dioxide and hydrogen. He discovered ten new ones, including oxygen. Priestley drew upon his domestic experience in investigating the effects of gases on living organisms. He used mice, in which his cottage abounded, keeping them in hygienically designed cages in a space behind the kitchen chimney, where the temperature was about 70 degrees Fahrenheit all the year round, because the fire was never allowed to go out.

The quantitative analytical methods developed by Black and the dazzling collection of new chemical facts discovered by Priestley were utilized by Antoine Laurent Lavoisier (1743–1794) to revolutionize chemistry and put it on a modern basis.

The great French chemist was primarily an intellectual organizer and administrator. He was neither a professor like Black nor a crafts-man like Priestley. He became one of the Farmers-General of taxes in France. These were private financiers who in the old régime under-took to pay agreed sums to the government in return for the right to collect the taxes. Most of them exploited the position to extract pri-vate fortunes from the taxpayers, and as a class they were fiercely hated. Their activities were one of the immediate causes of the French Revolution. Lavoisier was not one of the dishonest tax-farmers. He was efficient in collection and reasonable in his takings; but he shared the opprobrium of the class. There was a toughness in his character that made him disinclined to give anything away, and this came out in questions of priority in discovery. On several occasions, if he did not actually appropriate other men's discoveries, he did not vigorously protest when they were ascribed to him.

He became an outstanding industrial manager. He was appointed director of the French factory for making gunpowder. He improved gunpowder's explosive power and increased the output by a factor of five. His improvements were one of the causes of the subsequent victories of the French revolutionary armies. Lavoisier's scientific achievements were intimately connected with his military industrial work, for he was able to use the resources of the arsenal in carrying out his experiments. The chemistry of explosives was well suited to concentrate his attention on the nature of combustion.

He was a liberal in political outlook and sympathized with the original aims of the Revolution. He was, however, publicly identified with the hated tax-farmers, which led to his execution. The story that the president of the tribunal which tried him said that the 'Revolution has no need of scholars' is untrue.

The new interest in materials had led by the middle of the eigh-teenth century to the discovery and recognition of many new sub-stances, solid, liquid and gaseous. The differences between the various alkaline salts were recognized. Soda was distinguished from potash, and the alkalis from the alkaline earth, such as calcium and magnesium. Black's investigation of magnesia, which started from a consideration of its effects when used as a medicine, led him to identify the gas subsequently called carbon dioxide. He named it 'fixed air', conceiving it as a transformed kind of ordinary air. The chemists were still interpreting substances in terms of the four 'elements' of ancient times: earth, air, fire and water. Gases were generally regarded as varieties of ordinary air, which was supposed to be the elemental form of gas. Black's identification of 'fixed air' was of exceptional importance, because carbon dioxide has an extremely wide role in nature, being a product of combustion, of fermentation and of respiration. Black perceived these natural, industrial and biological implications.

In the same period there was a big increase in the knowledge of metals. Zinc was recognized as a specific substance, as were cobalt, nickel and bismuth. Platinum was brought from America in the middle of the eighteenth century. Its resistance to heat and its catalytic properties made it of great importance.

The flood of new facts led to intellectual confusion. Theories which had originally been proposed to explain a few facts became contradictory and broke down when applied to many of the spate of new facts.

The crucial chemical phenomenon at the beginning of the industrial era was combustion, the chemical changes in materials induced by heat. The German doctor and chemist, G. E. Stahl (1660–1734), brought some order into the confusion by propounding his phlogiston theory. This term, derived from the Greek for setting things on fire, was applied by Stahl to a weightless entity, which was supposed to cause substances that contained it to burn easily. The changes occurring when substances burned were due to the escape of phlogiston from them. Assuming that such an entity existed, it could be used to give a consistent account of a large range of phenomena. The conception was a modernized version of the ancient idea of the element of fire. The notion of an entity without weight did not seem unreasonable, as heat did not appear to possess weight, yet it had great potency.

Black's discovery of carbon dioxide, which was fundamentally different from ordinary air, was followed by Henry Cavendish's identification of hydrogen in 1765, and Priestley's discovery of oxygen in 1774. This made the ancient conception of ordinary air as one of the elements less plausible.

Priestley had discovered that ordinary air contains a constituent which supports combustion more strongly than ordinary air itself. He succeeded in producing this substance by heating red oxide of mercury, and demonstrated that a flame burned in it more brilliantly than in ordinary air. He interpreted the new substance as ordinary air which had lost its phlogiston, and he called it 'dephlogisticated air'. Cavendish then showed that water could be obtained by exploding together two volumes of his 'inflammable air' with one of Priestley's 'dephlogisticated air'.

The phlogiston theory gave a quite reasonable explanation of most of these experiments, but there were exceptions. In about 1771, when he was twenty-eight years old, Lavoisier had begun to study the phenomena of combustion. He soon formed the opinion that when a substance is burned in air it absorbs part of the air. He repeated the main experiments which had been made, and con- firmed the ancient observation, known since the time of Galen (AD 130–200), that when certain metals are heated in air their weight increases. This had been noted by various experimenters during the centuries, but Lavoisier applied to his experiments the quantitative analytical technique by means of weighing, which Black had in- vented in his experiments on the alkalis. He perfected his experiments until he constantly secured the same figures in his measurement of the increases in weight in combustion.

Lavoisier did not discover any new substances or new phenomena. His aim was different. It was to make experiments which would determine what happened in known phenomena, in order to decide whether one explanation of them, or another, was correct. These were what Bacon had called crucial experiments, because they decided whether a theory was incorrect. Lavoisier brought into chemistry the critical orderly spirit which he exercised with such great success in the collection of taxes and the organization of the manufacture of gunpowder. His outlook was different from that of Priestley and Cavendish, who were more interested in discovering new facts than new theories.

Black had demonstrated that the amount of 'fixed air', or carbon dioxide, absorbed by lime was exactly equal to the weight of the 'fixed air' which could be driven out of the resulting carbonate by heat. He explained this without invoking phlogiston. Lavoisier followed this by proving that when a metal was heated in a closed volume of air the increase in weight of the metal was exactly equal to the loss in weight of the enclosed air. His experiment was analogous to Black's, and it seemed to him that it should also be explicable without invoking phlogiston. At first he assumed that the part which the metal absorbed from the air was also Black's 'fixed air'. He did not grasp that it was oxygen until after Priestley, on a visit to Paris, had told him of his discovery of what he called dephlogisticated air, which caused a flame to burn more brilliantly than in ordinary air.

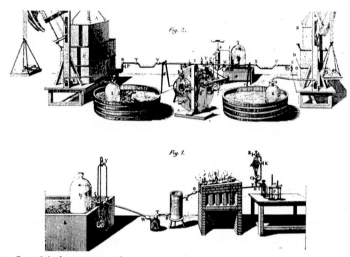

24. Lavoisier's apparatus for
the decomposition and recomposition of water

Lavoisier now began to perceive that the increase in weight of a metal when heated in air was due to combination with a part of the air which was fundamentally different from the rest. It was Priestley's new gas, which was not, as Priestley thought, air without phlogiston but another substance. At first Lavoisier called it 'vital air', and later *oxygen*, because water-solutions of compounds of it with metals were acid. He recognized that Cavendish's 'inflammable air' was also a specific substance, or element. He renamed it *hydrogen* (meaning water-forming) because the combustion of hydrogen in oxygen produced water. Lavoisier was the first to use the term 'element' effectively in the modern chemical sense. He began to restate familiar chemical reactions in terms of oxygen and hydrogen, without using the concept of phlogiston, which became superfluous.

He published his *Elementary Treatise on Chemistry* in 1789, in which the subject was re-formulated from this point of view. He enumerated thirty-three substances which, as far as contemporary knowledge went, appeared to be elements. Chemical reactions were expressed in quantitative terms after the manner of Black. This reduction of the subject to quantitative terms directed attention to the numerical relations between the precise amounts in which various elements combined with each other. The study of these relations by John Dalton (1766–1844) showed that many features of them could be explained on the supposition that the elements consisted of atoms, all the atoms of any particular element being identical in properties.

Black and particularly Priestley and Cavendish were individual workers. Priestley had the spirit of the ingenious craftsman, and Cavendish that of the talented amateur, but Lavoisier had added to his experimental ability a philosophical mind, which enabled him to introduce order into the new chemistry.

In his comparatively short life of fifty-one years he accomplished a great deal besides the revolution in chemical theory. With the collaboration of the great mathematician Laplace, he carried out remarkable quantitative researches on respiration. These were utilized by Robert Fulton in his first underwater dives with his submarine *Nautilus* in 1800, and acted as stimulus to the German medical doctor J. R. Mayer, which led him to the first published formulation of the theory of the conservation of energy. Lavoisier's systematic study of the chemistry of plant and animal substances, by himself and his colleagues, laid the foundation of organic chemistry, and before he died he had sketched out a programme for research on the chemistry of digestion. Lavoisier occupies in chemistry a place comparable with that of Newton in physics and Darwin in biology.

15. The interaction between industry, agriculture and science

When the Normans conquered England in 1066 they found in existence a self-contained system of communal agriculture. They did not make much alteration in the technical operation of this system, which remained substantially the same until the sixteenth century. Hitherto it had been conducted in the main for the subsistence of its practitioners. Now it began to be conducted for profit.

As each peasant worked many plots of land scattered over the district, he spent a great deal of time walking from plot to plot. Each plot was usually too small to be protected by hedges. Crops were often blown from one plot into another and became entangled. As a large fraction of plots were left fallow in order to recover from crop-bearing, weeds flourished on them and their seeds infested the cultivated plots. Thorough drainage, except under special conditions as in the Fens, became almost impossible, because of the small size and random distribution of the plots. It remained a difficult task until the invention and industrial production of the drain-pipe in the nineteenth century, and light motor-driven ditching machines in the twentieth.

Farming for profit, like a town business, stimulated men to look for efficiency. It was evident that the amalgamation of many small plots into larger units would save time and labour, reduce the amount of weeds and facilitate drainage. Acquisitive farmers bought up many small plots and made them into one considerable farm, which they then proceeded to clean, manure and drain more thoroughly than had been practicable under the old system.

This movement was a development of organization rather than technique. Scientific agriculture began in the seventeenth century, like the other aspects of modern science, and was a product of the same social outlook.

R. Weston (1591–1652), who had been in Holland as a Royalist exile during the Civil War, had noticed the Dutch cultivation of clover and turnips as field crops. Their adoption ultimately produced a revolution in English farming, which caused the great agriculturist Arthur Young to assert that Weston had been 'a greater benefactor than Newton' to mankind. The younger Lord Townshend (1674–1738), whose father had taken the initiative in inviting Charles II to return to England, had for his tutor the botanist William Sherard, who founded the chair of botany at Oxford. He and his tutor made a grand tour of Europe, and he returned a competent botanist. Townshend's botanical interest and knowledge enabled him to appreciate the value of turnips as a crop. He succeeded in introducing them on his estate as an alternative to leaving a third of the arable land fallow each year; this kept the land free from weeds. He further developed the system of rotation of crops, introducing the four-crop rotation consisting of turnips, barley, clover and wheat. The productivity of Townshend's estates greatly increased, and his rents went up by a factor of ten.

The increased production of crops created the conditions for a radical improvement of livestock. It was now possible to feed them properly in winter and keep them alive for a long period; and the enclosures kept them under control, so that they were no longer all mixed together and breeding promiscuously. Scientific selection for the improvement of livestock became possible.

The improvement of the mechanics of agriculture began at the same time as the biological improvements. The ancient implements, such as the plough, harrow and scythe, had not been fundamentally improved for more than a thousand years. They were tools rather than machines, with no moving parts. The fields resulting from enclosure presented larger areas for uniform working. The ancient hand implements were more suited to working the small and variable old plots. With the rise of agriculture for profit, men began to seek more versatile equipment. One of the first processes to receive attention was that of sowing. Was it not possible to devise a machine which would plant corn evenly, saving labour and producing more

K

uniform growth? Christopher Wren in his inventive youth applied himself to the problem.

The first success was gained in 1700 by the country gentleman Jethro Tull (1674–1741). He had been much impressed with French methods of cultivating the vine, and had observed the beneficial effects of regularity of planting and of continual agitation of the top-soil by hoeing and ploughing to remove weeds. This led him to conceive that seeds of corn should be planted evenly in straight rows in well-tilled ground, like vines. He personally planted corn on this plan in his garden and got better results, but when he tried to introduce it in his fields he failed, because his labourers either could not or would not learn the new method. He therefore decided to try to make a machine which would plant the seed as he wanted.

After many experiments, he devised one which would set seed at constant spacing, independent of the speed at which the machine travelled. It sowed the seed in straight rows, leaving a space between rows which could be weeded and hoed. He devised a hoe hauled by a horse for this purpose. Tull's crops produced three times the average yield. He believed that tilth, or soil reduced to a fine condition by much tilling, was more important than manure. He grew wheat successively for twenty-three years on the same piece of soil, without manuring, getting bigger crops than farmers who used manure and traditional methods of cultivation.

The adoption of Tull's seed drill was at first slow. Labourers refused to use it because it made many of them redundant. They disliked the machines because these often broke down, mechanical engineering not having advanced sufficiently to make them reliable. The development of agricultural machinery remained slow until the Industrial Revolution improved engineering, and the rapid increase in population created a still more insistent demand for food.

The new urban population could not perform heavy industrial work without a meat diet. This led to a big development in animal breeding. The Leicestershire farmer Robert Bakewell (1725–1795) produced a new breed of sheep which gave more meat in proportion to bone, and twice as much meat as the traditional breeds. He obtained his results by systematic inbreeding, that is, by mating related animals, in such a way that their best points were emphasized. His methods laid the foundation of the British pedigree herds, which have had great effect in raising animal productivity in many parts of the world.

The enclosure of land also made possible the selection of the best crop plants. A big advance occurred in 1820, when an agricultural labourer called John Andrews noticed a giant grain of barley roll out of his boot after he had come home from harvesting. He planted it in the following spring and obtained a harvest of grains of similar size. A local clergyman, the Reverend John Chevallier, heard of it and began to cultivate it. The new barley became famous under the name of 'Chevallier' barley; it was not yet thought proper for a new crop plant to be named after a mere agricultural labourer. In recent times barley has been further improved, especially by big brewing firms whose scientists conduct extensive research on the selection of the best kinds of grain for brewing. The improvement of wheat by selection has greatly increased the food resources of the world. As Lamartine Yates has said: 'Yields of wheat, which had been static from Nero to Napoleon at around ten bushels an acre, had risen to fifteen bushels by 1850, and averaged twenty to thirty bushels in some European countries by 1900. Today, yields in some countries exceed fifty bushels per acre.' Under the best conditions, yields are much higher still. The improvement of wheat coincided with the development of the Industrial Revolution.

A big stimulus to the application of chemistry to agriculture arose from the invitation to Humphry Davy to lecture on this subject in 1803, when the food shortage was intense owing to the Napoleonic wars. About a quarter of a century later the German chemist Justus von Liebig carried it forward with wonderful power. He had invented methods of analysing plant and animal materials, or organic compounds, which were sixty times as fast as those previously in use. With them he obtained a great deal of new knowledge in a short time.

They enabled him to trace particular chemical substances, such as certain salts, through the whole cycle of life, from their absorption from the soil by the plant and thence into the tissues of the animal which fed on the plants. This led him to perceive that these salts were necessary for life. They were an essential constituent of natural manures. Liebig contended that they would be equally effective if fed in their pure form from any other source. As a result of Liebig's suggestions, the nitrates found in great beds in Chile, formed from desiccated droppings of countless millions of seabirds in the past, were imported into Europe and used as manures, constituting a whole new industry. He arrived at the idea of artificial fertilizers which could be made by chemistry.

The technical problem of making and using such a fertilizer was first solved satisfactorily by J. B. Lawes and J. H. Gilbert. They had studied chemistry, and Gilbert had been one of Liebig's students. They were aware that phosphates could be made soluble by acid treatment, and it occurred to them that bones dissolved in acid would be much more assimilable by plants. The discovery of Lawes and Gilbert enabled agricultural soil, exhausted in many places in Britain and Europe by centuries of cropping without adequate manuring, to support a large part of the increase in population during the nineteenth century. With the fortune that he had made from artificial phosphate fertilizers, Lawes founded the Rothamsted Agricultural Research Station.

Liebig had recognized the importance of nitrogen for plant growth. He suspected that it was obtained by the plants from the air but was unable to find out how. The clue was discovered by Schloesing and Muntz in 1877, who were investigating the process of sewage purification. They found that it was due to the production of nitrates in the sewage. This did not happen quickly but slowly, as if it were the product of a living process. They argued that if the sewage were alive it ought to become quiescent when chloroformed. They tried the experiment and found that this did indeed happen. A bacteriologist then showed that the living organisms in sewage were bacteria. It followed that the nitrates found in ordinary manure were made from the nitrogen in the air by bacteria.

After this it was discovered that the nodules on the roots of leguminous plants, such as clover and peas, contained bacteria which could fix nitrogen from the air. This was one of the reasons why clover served such a valuable purpose in the rotation of crops. The next step was to try to get nitrogen directly from the air by chemical means, and then feed it in the form of synthetic nitrates to the soil. This was first done successfully in Norway by K. Birkeland and

S. Eyde, by blowing air through an electric arc which is intensely hot. This caused some of the nitrogen and oxygen in the air to combine. The substances so formed could later be dissolved in water and converted into nitrates.

The electric-arc process consumed a great deal of electricity, and was superseded by the process worked out by F. Haber (1868–1934) in 1913, of combining nitrogen and hydrogen by means of catalysts to produce ammonia, from which nitrates are easily made. The production of synthetic fertilizers was enormously increased during the Second World War. The world output of synthetic nitrates rose to a quantity which contained the equivalent of four million tons of nitrogen from the air. About six million tons of phosphates were extracted from phosphate rocks.

After the simpler chemical constituents of plants and animals had been recognized, attention was devoted to the more subtle chemical constituents of living things. This led to the discovery of vitamins and hormones. Vitamins were first recognized by Eykman in Indonesia, and F. Gowland Hopkins in England. J. F. Eykman (1851–1915) had noted in 1890 that beri-beri, a disease characterized by anaemia and a general lack of health, was caused by eating polished rice. He showed that the bran obtained when rice was polished contained a substance soluble in water and alcohol, which could prevent beri-beri. By 1912 Hopkins had proved conclusively that certain substances were necessary, though only in very small quantities, for normal growth and well-being in rats. He called these 'accessory food factors'. His precise description was gradually superseded by the less accurate but more picturesque term: 'vitamin'. The growth-promoting hormones in plants were discovered by F. A. F. C. and F. W. Went in Indonesia. They were first synthesized chemically by F. Kögl in Holland.

As advances in chemistry in the early part of the nineteenth century led to a deeper understanding of the needs of plants, and to the foundation of artificial fertilizer industries, so advances in the twentieth century have led to the development of new industries manufacturing a wide range of subtle chemical substances, which promote growth, affect the behaviour of plants – such as the setting of fruit – kill insect pests and destroy weeds. The rapid increase of the world's population is one of the factors that stimulates the progress of these developments. Chemists search for new substances which may be of agricultural value, and this demand stimulates efforts to improve chemical methods. As Liebig radically advanced methods of organic chemistry in his day, so contemporary chemists have developed such new techniques as chromatography, invented by M. Tswett in 1906, and developed by A. J. P. Martin and R. L. M. Synge in 1941. These techniques have greatly increased the power of analysing and synthesizing the complicated molecules in living substances.

The combined effects of the various applications of science to agriculture have been great, but the most potent single factor has probably been the tractor driven by an internal-combustion engine. In its latest forms, with hydraulically operated mechanical attachments, it extends human powers of manipulation and is far more than a source of power. In 1939 horses outnumbered tractors in Britain by 13 to 1. Today, horses have virtually disappeared from agriculture. The tractor operates quickly and has helped to emancipate the farmer from the weather.

In addition to improving agriculture, science offers the ultimate possibility of synthesizing food from minerals. Some progress has been made in synthesizing edible fats from petroleum. As the demand for food becomes ever more insistent, and as science advances, there is little doubt that ways of synthesizing food on a large scale will be worked out.

16. Combatting diseases new and old

The explorers of the New World returned with new diseases as well as new materials and new species of plants and animals. The conventional system of medicine, inherited from Galen and based on centuries of Old World experience, failed to cope with the ravages of syphilis, imported from Mexico. The only effective medicines against this new disease were found to be chemicals, not of plant or animal but of mineral origin. Under the leadership of Paracelsus (1493-1541), this application of chemical knowledge of industrial origin to medical chemistry stimulated great advances in chemistry as a whole, besides producing successful innovations in medical treatment.

Paracelsus was, however, a fantastic personality, besides being a man of genius. He behaved like a wizard and magician, and created the impression that the fresh impetus which he gave to chemistry was of alchemical inspiration. However, as Daumas has pointed out, 'Despite the widespread opinion to the contrary, technicians rather than alchemists laid the foundations of modern chemistry.' Long before the dawn of history, man was using processes involving the oxidation and reduction of metals, though of course he did not conceive his operations in these modern terms. He used fermentation for preparing food and drinks and making hides usable for clothing. Neanderthal man dug up manganese oxide to use it as pigment. The invention of textiles led to the development of dyeing, the spread of knowledge of which perhaps did more than anything else to promote early chemistry. Considerable knowledge of such techniques as dyeing and gilding was accumulated over thousands of years. The technique of gilding was particularly stimulating, for it involved a variety of chemical procedures with metals. Already in Alexandrian times there were well-established recipes for gilding, in which arsenic compounds had an important place.

Paracelsus drew upon technical chemistry, both old and new, and applied it in a new way, thus opening up new scientific perspectives. His original name was Philippus Aureolus Theophrastus Bombast von Hohenheim. He was born in the formative time after the discovery of America. His life and work were an expression of one aspect of the deep forces stirring in Europe which had inspired that crucial achievement. He was the son of a professor at the School of Mines in Southern Austria. He acquired his basic knowledge of chemistry from mining and had experience in work underground. He became interested in medicine, listening to lectures at many universities and seeking all the doctors, alchemists, astrologers and magicians he could find, to learn of secret and new cures and remedies.

He heard of the terrible diseases from the New World and the failure of traditional Galenist medicine to deal with them. He was thrilled to learn that only minerals, which had been his first youthful enthusiasm, were effective in treating them. The Galenist remedies consisted of mild extracts from plants and animals, associated with careful regulation of diet and regimen. Paracelsus therefore conceived the need for a new medicine based on violent drugs made from the minerals which first fascinated him. His ambitious, masterful, propagandizing temperament made him an effective instrument of the transforming forces of the age, and he became possessed with the determination to sweep away the traditional medicine and establish a new one, based in particular on chemicals of mineral origin, such as mercury and antimony.

His mineral drugs did indeed effect some cures for which traditional drugs were useless. This was a foundation on which he was able to secure popular support through his propagandist genius. He generated such excitement that the authorities were forced to appoint him professor of medicine at Basle in Switzerland in 1526. He started his course by collecting all the traditional textbooks on medicine, piling them up before his pupils and setting fire to the lot. He told his followers to ignore the books and study nature directly, especially the properties of metals and minerals, in order to discover new cures and remedies.

He brought industrial ideas, conceptions and ways of doing things into medicine. Through this, he helped to emancipate medicine from the ancient traditions of magic and actually undermined alchemy, in spite of his wild behaviour. He presented his ideas in obscure verbiage, and lived in a continual uproar of controversy and abuse. His own name of Bombast became a universal term for boasting. Nevertheless, he more than any other man started the new era of chemistry.

Largely as a result of Paracelsus's influence, chemical medicine rose to a commanding position in the seventeenth century, under such leaders as H. Boerhaave (1668–1738) at Leiden, whose lectures were attended by chemists from many parts of Europe, and especially from Scotland. It was this development which caused Liebig to remark that 'the great physicians, who lived towards the end of the seventeenth century, were the founders of chemistry'. Since Paracelsus, the contribution of chemistry to medicine has continued with unabated power.

Priestley breathed oxygen soon after he had discovered it, and noted its physiological effects. This started numerous researches on the medical properties of gases, and led to the young Humphry Davy's discovery of the anaesthetic properties of nitrous oxide, another of the gases discovered by Priestley. Then the anaesthetic properties of ether and chloroform were discovered.

The development of the dyestuffs chemistry in the nineteenth century provided the technique for the synthesis of an ever-extending series of anaesthetics and drugs, typified by the synthesis and industrial production of acetyl salicylic acid, or *aspirin*.

Yet again, a chemist was to have a revolutionary impact on medicine. This was Louis Pasteur (1822–1895), who frequently remarked: 'I am ignorant of medicine and surgery.' The only medical degrees he received were honorary. Pasteur's first research was on crystals. He discovered that common tartaric acid consisted of right-handed crystals only, while a rare form of the acid, found in wine-casks, consisted of equal quantities of right- and left-handed crystals. It appeared that living organisms worked only with right-handed crystals. For some reason living processes operated in a chemically lop-sided way; instead of operating with equal numbers of right- and left-handed crystals, they built up living systems with only right-handed crystals. There was something essentially asymmetrical at the heart of nature.

This discovery still reverberates through biology and the latest knowledge of the inner structure of living matter. Pasteur suspected that cosmic forces, operating from outside the earth, exercised some

25. Pasteur's type of long-necked flask

form of selection on the kind of molecules which could be utilized in the process of growth by living organisms. He mixed various minerals together and exposed them to strong magnetic fields, in imitation of the conditions that might have existed when living matter was first formed on the earth. His experiments gave no result but were very modern in spirit. In recent years some of the molecules out of which proteins are formed have been successfully synthesized from a mixture of hydrogen, water-vapour, ammonia and methane, agitated by electric sparks simulating lightning flashes or discharges, such as might have occurred on the earth 2,000 million years ago, when living organisms first appeared.

Pasteur's remarkable work on crystals established his scientific reputation. After some years he was appointed professor at Lille, in northern France, where he was expected to apply chemistry to local industries. One of these was brewing, and so he began to study fermentation. He soon announced that 'fermentation is essentially a correlative phenomenon of a vital act beginning and ending with it'. It did not occur without the multiplication of living globules. In 1857 he was called to Paris as professor. There he continued his researches on living globules, or micro-organisms, and worked out the technique of pure cultures, by which the various kinds of micro-organisms could be distinguished. He became involved in controversies on whether life could be generated spontaneously, and proved conclusively that all alleged experiments purporting to demonstrate this were fallacious; so far as was then known, life could arise only from life.

In 1862 he pointed out that the study of micro-organisms constituted a first step in the serious investigation of contagious diseases. Then he was asked by his fellow-townsmen why their wine had gone sour. He identified the micro-organism which had caused the trouble, and showed that if the wine were heated to 60 degrees Centigrade most of the micro-organisms would be killed, and the wine would keep. Thus he invented 'Pasteurization'. After this, he was asked to investigate the disease which had decimated the worms which produced the silk for the French silk industry. He had never even seen a silkworm cocoon. One was handed to him; he shook it near his ear and remarked: 'It rattles: there is something inside.' From this stage of knowledge of the problem he embarked on a tremendous programme of research that lasted years. The life history of the micro-organism which was causing the silkworm disease was very complicated, but Pasteur stuck to the study of it, together with all his routine work. His efforts were such that in 1868 he had a stroke and thenceforth was slightly paralysed, but it did not impair his intellectual energy. It did, however, affect his method of work. He became dependent on assistants for experimental manipulations, and he began to devote himself more to the intellectual organization of discovery. He made detailed card-indexes of everything bearing on his problems, and spent hours, one, two, three, four, five ... sitting motionless in a 'brown study', meditating on the entries on his cards. No one dare interrupt him at these times, and all walked around on tiptoe. In the year after his stroke he solved the silkworm problem, identifying the micro-organism and providing instructions on how it was to be evaded. Thereby the French silk industry was saved.

During the Franco-German War he had a passionate feeling against the Germans. After the war he conceived the plan of striking at the German beer monopoly by finding out how to make French beer as good or better. He made outstanding researches on how to cultivate pure yeasts by which the problem could be solved.

Pasteur was now asked to investigate the anthrax disease which was ravaging French sheep. It was already known that the blood of infected animals was full of small threadlike bodies, and Koch had shown that they could be cultivated outside the animal. The veterinary doctors could not believe that these bodies were the cause of the disease, for years after the disease had apparently disappeared from a district, it suddenly reappeared. The cause could not be a living organism that had been there all the time. Pasteur showed that the organism retained its virulence, even after being bred through a hundred generations. He explained that when anthrax-infected animals were buried the spores of the micro-organism, which was one of those that multiplied without free oxygen, remained alive, and some were ultimately brought to the surface by earth-worms.

After having elucidated the anthrax problem he began to study human diseases, to which he applied the bacteriological techniques that he had worked out for dealing with the anthrax organism.

Meanwhile, he investigated fowl cholera. In the course of this work he found that most of his fowl had died while he was away on holiday. He took specimens from the organisms of the remaining dying fowls, cultivated them and then injected doses into healthy fowls, in order to secure a new flock of infected birds. The healthy fowls remained well, and so he then tried injecting them with fowl cholera from a new source. To his surprise, they continued well. The injections with the old culture had made the birds immune to new infection by the disease. This was the beginning of the science of immunology.

He succeeded in making preparations of anthrax which, when injected into healthy sheep, made them immune to infection by fresh anthrax. After this he succeeded in preparing a vaccine against rabies; yet another major advance, for rabies is produced by a virus far smaller than a bacterium, so small that it could pass through a filter, and too small to be seen by an optical microscope. The convulsions produced by rabies are caused by the virus attacking the brain and the spinal cord. Pasteur therefore cultivated the virus in the brains of rabbits, and succeeded in producing a weakened virus which could be used as a vaccine against rabies.

Pasteur's development of bacteriology suggested to Joseph Lister (1827–1912) the idea of using antiseptics to kill the infectious organisms which had so often entered the wounds of persons undergoing operations. Lister's antiseptics were effective outside the body, but had drawbacks when used in wounds, where they damaged healthy tissues as well as attacking the germs.

The first adequate success in using chemicals to kill bacteria within the body was achieved by Paul Ehrlich (1854–1915) in 1909. He tried the effect of numerous substances on the syphilis organism. The six hundred and sixth which he tried, a compound of arsenic, proved successful. This substance, known as 606, or *salvarsan*, had the remarkable property of attacking the syphilis organism and nothing else.

Attempts to discover other chemicals effective against other organisms advanced slowly. In 1914 Eisenberg noticed that an azo dye, which is a compound of aniline containing 'azote' (nitrogen), would kill certain microbes. In 1930 the German chemical industry started a systematic investigation of the bacteriocidal properties of this class of dyes, and three years later Domagk (1895–1964) published his discovery that the azo dye prontosil was effective against many bacteria. The killing action was due to a particular part of the molecule: para-amino-benzene sulphonamide. This was the start of a new chemical revolution in medicine. Sulphonamides proved effective against childbirth fever, venereal diseases, cerebrospinal meningitis and pneumonia.

The sulphonamides were not, however, effective against bloodpoisoning. The threat of a second world war caused anxious attention to be devoted to searches for drugs which would prevent the terrible destruction from blood-poisoning which occurred in the First World War. This strengthened interest in H. W. Florey's researches on the problem of natural immunity. In the course of his investigation, he studied the behaviour of two antibacterial substances discovered by Alexander Fleming (1881–1955). One of these was lysosyme from human tears, and the other penicillin. Fleming had found penicillin unstable, and therefore not of much practical value in the form in which he had obtained it. Florey, Chain and their colleagues showed that the active principle was an organic acid, and worked out methods for making stable preparations of it, thus converting it into a practical drug. It was the first of the antibiotics.

The sulphonamides and antibiotics have transformed modern medicine. Their action is described as bacteriostatic rather than bacteriocidal, for they prevent the bacteria from growing and multiplying; the bacteria consequently decay and pass harmlessly away.

17. Electricity

The success of Paracelsus in applying mineral chemicals to the cure of disease directed more attention to the curative effects of other agents of non-living origin.

The ancient Greeks noted that when amber, a kind of fossilized resin, is rubbed it attracts feathers. The Romans observed that a certain kind of stone found in the Italian district of Magnesia attracted pieces of iron, and that iron rubbed with such a stone might acquire the same power. The Romans were also aware that the torpedo fish could give a shock if touched, and their doctors even prescribed shocks given in this way for the treatment of gout. They had no insight into the nature of the shock. Lightning had been known and feared by man ever since he came into existence.

All of these phenomena were probably exploited by medicine-men before the dawn of history in performing magical rites. When David Livingstone made his explorations in Africa in the middle of the nineteenth century he found primitive tribes who were familiar with the electrical effects produced by rubbing furs. Medicine-men supposed that substances with attractive properties might help a girl who had been jilted to regain the affection of her former lover. She might do this by holding a piece of magnetic stone in her hands.

The first thorough review of the materials with attractive powers which are found in nature was undertaken by William Gilbert (1540–1603) before the year 1600. He approached them from a strictly naturalistic, modern point of view, dismissing all stories of their magical powers as 'idle tales and trumpery'. Reference has already been made to his work on magnetism. While exploring magnetic attractions, he studied the somewhat similar attractions exerted by rubbed amber. He extended the investigation to other substances, and found that glass, resin, sulphur, diamonds and sapphires could also be made attractive by rubbing. He described all these substances as 'electrics', deriving the name from the Greek name for amber, *elektron*, or 'bright'. The word 'electricity' is derived from Gilbert's 'electric'. Gilbert did not discover that metals could be electrified by friction, because he held the piece to be rubbed in his hand, with the consequence that the electricity flowed off the metal through his body to the earth. He discovered, however, that electrified bodies could be discharged by flames and by damp; both phenomena of fundamental importance.

The next major advance was made by Otto von Guericke (1602–1686), the mayor of the German city of Magdeburg and organizer of military supplies to King Gustavus Adolphus, the Swedish Protestant conqueror. Guericke studied medical science in Holland. His greatest contribution was the invention of the air pump (Fig. 10). He thought always in engineering terms, and so it was in keeping with this that he also invented the first electrical machine. It consisted of a sphere of sulphur which could be rotated on an axle by an assistant, while a charge was accumulated on the sphere by holding the hand against it. Much bigger charges were obtained from this manual-power-driven machine than from rubbing a piece of material held in one hand by the other. Guericke discovered that an electric charge might repel as well as attract. He observed that when he brought his finger near the charged sulphur sphere a crackling could be heard and sometimes a luminosity seen. Leibniz followed this up, and proved that electricity could produce sparks.

In 1706 Wall compared the crackling and flashes from large pieces of amber with thunder and lightning. This was the first suggestion that lightning was electrical.

Considerable electrical research now proceeded. Stephen Gray (1666–1736) showed that electrical effects could be sent through a cotton thread, 866 feet long, suspended from poles by loops of silk. It was in principle an electric telegraph. He discovered the distinction between conductors and non-conductors, and even showed that water could be electrified, by electrifying soap bubbles. Gray observed that pencils of light could be seen to stream out of charged metals when looked at in the dark. This was the laboratory discovery of the brush discharge, which is seen during thunderstorms from the tops of masts on ships or buildings, and is known as Saint Elmo's Fire.

F. Hawkesbee made an improved electric machine, substituting a hollow glass sphere for Guericke's ball of sulphur. He published in 1709 the observation that if the glass sphere were exhausted with an air pump, and the sphere were then electrified, a glow would appear inside the sphere. This was the discovery of the principle of the electric discharge lamp. Already in 1744 Grummert in Germany proposed the use of vacuum electric discharge tubes as lamps in mines, and spelled out the name of King Augustus in such lamps. In 1752 Watson made a discharge tube thirty-two inches long that gave a steady light. Meanwhile in 1745 Musschenbroek in Holland invented the Leyden jar, or electrical condenser, in which large electric charges could be stored. With such condensers, powerful electric shocks could be produced.

So far, the study of electricity had been descriptive. More precise thinking about its nature was advanced by Benjamin Franklin (1706–1790), the first great scientist born in and identified with America. By about the age of forty he had made a modest fortune from publishing, which gave him leisure to pursue his interest in science. He started to experiment with a set of electrical apparatus sent to Philadelphia, and attempted to make a clear theory of what he observed. He concluded that electricity was not created by friction, but was 'really an element diffused among, and attracted by other matter'. He described the two kinds of electricity as positive and negative, and represented them by + and − signs. He showed that electricity consisted of a thing which could come and go but never be destroyed. It existed in definite quantities, which could be the subject of mathematical calculation.

He further showed that the power which a Leyden jar had of giving a shock resided 'in the *glass* itself'. This fundamental observation was the forerunner of Faraday's discovery that the seat of electromagnetic action is in the space around the conductor, which ultimately led to the discovery of radio waves. Franklin introduced the term *battery* to describe a series of Leyden jars joined together to increase the discharge. He even made a small electric motor which would run for half an hour on the electricity stored in a Leyden battery.

L

He conceived electricity as extremely subtle particles which could move through the densest metals with no perceptible resistance. He explained the fan-shape of the brush electric discharge as due to the mutual repulsion of the electric particles which constituted it. His study of the brush discharge from sharp metal points of charged conductors led him to invent the lightning conductor, which created an enormous public impression. Besides being of practical use in protecting buildings, and especially stores of gunpowder, the invention brought a force of nature, feared since the advent of man, under a degree of human control. In his famous kite experiment, in which he drew electricity from a thundercloud, he showed that such clouds were usually negatively charged. This remained the only precise piece of electrical information on thunderstorms for 170 years.

The spectacular effects of electric shocks on animals and men led to a great deal of experimenting on their application in medical treatment. Persons suffering from paralysis were subjected to shocks from electricity brought down from thunderclouds, as well as by shocks from Leyden jars and electrical machines.

Attempts were made to discover exactly how electricity exerted its effects on living organisms. Among these experimenters was the Italian anatomist Luigi Galvani (1737–1798) of Bologna. He had been interested in the way in which the nerves control the body. He stimulated the nerves in frogs' legs by pressing them with a knife; the nerves then caused muscular contractions. In 1780 a man happened to be working an electrical machine in his laboratory. It was noticed that while the machine was being operated a very light touch with the knife on a nerve, which would otherwise have produced no effect, now caused a large kick in the frog's leg. Galvani seized on this fundamental observation and investigated it tenaciously for eleven years. He discovered that electricity from a lightning conductor produced twitching in a frog's leg. Then he found that if the muscle of the frog's leg were laid on an iron fence in his garden, and the nerve which controlled it fastened to the fence with a brass hook, the leg twitched. He concluded that the electricity came from the animal tissue, and so described it as 'animal electricity'.

Galvani's experiments attracted the attention of Alessandro Volta

(1745–1827). He concentrated on the physical aspect of the phenomena, on the electricity rather than the animals. He devised more
sensitive instruments and used these to analyse what Galvani had
observed. He presently discovered that the electricity did not come
from the animals but from the metals. The twitching frog's legs were

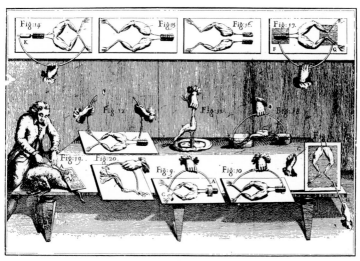

26. Galvani's experiments on frog's legs

merely acting as detectors of electricity which apparently arose from
two pieces of different metal when in contact.

   He put a piece of tin foil on the tip of his tongue and a silver coin
on the back, and found that when he joined the foil and the coin
with a wire he could detect a constant sour taste. He was using himself as a detecting instrument. He could tell the direction of the electrical effect from the position of the taste. He had identified the steady
electric current. He soon reproduced the set-up in his mouth in the
shape of pieces of zinc and copper separated by wads of cloth soaked
in dilute acid. Then he made a pile of these, so that the electric
current could be made more intense. Thus he invented the Voltaic
Pile. He sent a description of it to the Royal Society in London,
which published it in 1800. In the meantime, news of it went

round London, and even before its publication Nicholson and Carlisle had discovered that the electric current from the pile could decompose water.

In 1801 the young Humphry Davy (1778–1829), a Cornish boy whose scientific ability had been noticed early, came to London. James Watt's son Gregory, a talented chemist, had lodgings with Mrs Davy, the widowed mother of Humphry, and became friends with him. The Watts recommended Davy to Dr Beddoes of Bristol, who, inspired by Priestley's discoveries, had been experimenting with the medical possibilities of the respiration of gases.

Davy became his assistant, and launched into original researches on the properties of nitrous oxide. He discovered its property of making people who inhaled it laugh, and found that it could temporarily remove the pain of an aching tooth, thus discovering anaesthetics. Within eighteen months he had made his scientific reputation. After this Davy was engaged by the Royal Institution in London. He followed up the new electrical researches, and used a Voltaic battery to decompose potash and soda, hitherto regarded as elements. As a result of the decompositions, he discovered the astonishing new metals potassium and sodium. Then he discovered the electric arc light and the electric arc furnace.

Davy's discoveries showed the intimate relation between electricity and matter. It was evident that chemical affinity must be electrical. He saw that certain minerals must produce electric currents in the earth, and suggested that minerals should be located from such currents. This technique of electrical prospecting is now widely used in searching for oil and other minerals. He pointed out that the electric current, being able to transport chemical substances, might be used to remove damaging substances from the human body. This has now been developed for medical purposes.

The next fundamental electrical advance came with H. C. Ørsted's (1777–1851) discovery in 1819 that an electric current could move a magnetic needle. Scientists everywhere tried to use the new effect to obtain continuous rotation from an electric current, that is, to invent an electric motor. The first to succeed in this was Davy's laboratory assistant, Michael Faraday (1791–1867), who demonstrated his 'electromagnetic rotations' in 1821.

Now that magnetism had been obtained from electricity, the scientists of the world tried to obtain electricity from magnetism. Then, in 1831, Faraday showed that an electric current was produced in a

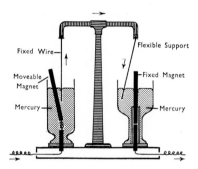

27. Faraday's electromagnetic rotations

wire coil when a magnet was pulled out of it suddenly. He had discovered that *relative motion* between coil and magnet was necessary for the production of the current. It had been difficult to see this in Ørsted's experiment, because the magnetic needle stayed in a steady position as long as the current in the wire remained steady. Everything that could be seen by the eye was stationary, which had caused experimenters to fail to note that the *current*, which they could not see, was in motion. The role of relative motion in producing electromagnetic effects ultimately led to the discovery of the theory of relativity.

Faraday now embarked on an equally far-reaching series of experiments on the conduction of electricity in liquids. This led him to the precise definition of the subject, into which, with the help of William Whewell (1794–1866), he introduced the terms *electrolysis, electrolyte, electrode, anode* and *cathode*; now all household names. He introduced the word *ion* to describe the electrified substances that were transported through liquids in electrolysis. He discovered the exact proportions in which various elements were set free in electrolytic decomposition by a standard current, and showed that the amount of each element set free was associated with a precise and constant quantity of electricity. Forty-seven years later H. Helmholtz (1821–1894) pointed out that this quantity was equal to the charge of an electron. Faraday had discovered the ultimate unit of electricity, but he was loth to recognize it as an *atom* of electricity, for, like Davy,

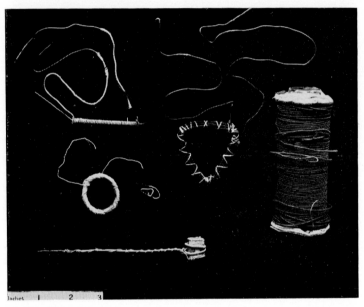

28. Faraday's coils

he said that he was 'jealous of the term *atom*, for though it is very easy to talk of atoms, it is very difficult to form a clear idea of their nature'.

Faraday's electromagnetic discoveries were taken up by James Clerk Maxwell (1831–1879) in 1855. He saw that Faraday's conception for describing the interactions of electricity and magnetism was susceptible of precise mathematical description. Maxwell succeeded in formulating it in terms of mathematical equations. These equations indicated that waves similar in character to light-waves, but different in wavelength, might exist. H. Hertz (1857–1894) demonstrated in 1887 that these waves, radio waves, could actually exist.

Meanwhile, Maxwell had pointed out that as light consisted of electromagnetic waves travelling at a certain speed, it was important to discover what effect the speed of a source of light had on the speed of light emitted by it. In 1881 the American scientists A. A. Michelson (1852–1931) and E. W. Morley (1838–1923) made a decisive experiment, which proved the apparently paradoxical result that the speed of the source had no effect on the speed of the light emitted from it. At last, in 1905, Albert Einstein (1879–1955) discovered how this paradox could be resolved, but it required a fundamental revision of the traditional Newtonian conceptions of space and time.

While Faraday's electromagnetic discoveries had opened the way to the discovery of radio and relativity, his researches on electrolysis were equally fruitful in a complementary direction. After his study of the passage of electricity through liquids, he started on the study of its passage through gases. Following his line, J. Plücker (1801–1868) in 1858 discovered what are now called cathode rays. Considerable but inconclusive research was done on them during the next twenty years, and then the physicists' attention was diverted by Hertz's great discovery of radio waves in 1887. Hertz himself began to investigate the cathode rays, and found they could go through metal foils. He doubted that charged particles could go through metals, and concluded that cathode rays must be waves.

J. J. Thomson (1856–1940) showed in 1894 that the speed of the cathode rays was only about one-twentieth that of light, which indicated that they were almost certainly not electromagnetic waves. Then, in 1895, came the revolutionary discovery of X-rays. K. Röntgen (1845–1923) at Würzburg noticed that some of his unused photographic plates were fogged. There was no reason to suppose that they were bad plates, and so he searched for the cause. He was thunderstruck to find that something was emitted by his electric discharge tubes, which penetrated the dark wrappings of the unused plates and fogged them. Röntgen examined every aspect of the phenomenon in excitement and secrecy, not even telling his wife. Within a few weeks he had made a clean sweep of the fundamental properties of X-rays. Among these was their power of making air electrically conductive.

J. J. Thomson seized on this. It was just what he was wanting for his experiments. He and his pupil E. Rutherford (1871–1937) found that they could easily make the gas in their discharge tubes conducting at a mere 300 volts with X-rays. This greatly facilitated the experiments, and by 1897 Thomson had proved that cathode rays consisted of electrically charged particles of about one-thousandth

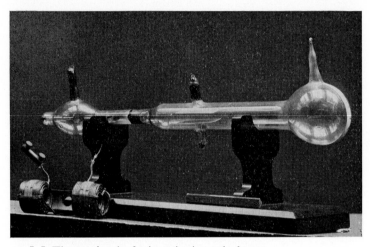

29. J. J. Thomson's tube for investigating cathode rays, which led to the discovery of the electron

the mass of the hydrogen atom. As these particles, or electrons, could be obtained from any element, they were evidently parts of the atoms of such elements, and it followed that the atoms of the elements must have a structure. Thus the age of electronics and atomic structure began.

The discovery of X-rays prompted the search for similar rays. The French mineralogist H. Becquerel (1852–1909), on the prompting of the mathematician Henri Poincaré, investigated his collection of minerals, and in 1896 announced his discovery that uranium minerals emitted a radiation which, like X-rays, could fog photographic plates through their wrappings. He had discovered radioactivity.

Pursuing this discovery, Pierre (1859–1906) and Marie (1867–1934) Curie found that uranium minerals contained a fraction which was immensely more radioactive than uranium itself. In 1898 they announced their discovery of radium.

The twentieth century was now at hand.

18. The theory of energy

The theory of mechanics worked out by Galileo and Newton was devised to describe the movements of bodies which are subject to negligible amounts of friction, such as planets moving in space around the sun or balls running down very smooth planes. In Newtonian mechanics the mathematical function consisting of the product of force and distance, or 'work', was found to be useful in solving problems of frictionless motion. This notion did not come from the study of planets in motion, but of objects on the earth moving against a resistance.

The 'work' done by the moon when it fell through a certain distance was an abstract conception. The 'work' done when a quantity of coal was raised out of a mine was expressed in exactly the same mathematical symbolism, but its content had an extremely different nature. It cost effort and had a financial value. Consequently the attitude of mathematical astronomers and engineers to the same symbolism was quite different. For astronomers the function consisting of one-half the mass of a body multiplied by the square of its velocity was useful in solving equations of motion; for engineers it was a measure of 'accumulated work', which, when done, would have a commercial value.

The full scope of the notion of energy was not explored until commercial interests made it necessary. This came after the development of the steam engine, when it became essential to understand the relation between the output of power and the amount of fuel consumed. Fuel consumption was a prime factor in the cost of the power used by the manufacturer. There was therefore a motive for measuring the performance of a power-plant and trying to make it more efficient.

The first exact measurements of heat made by Joseph Black and others seemed to them best explained on the theory that heat con-

sisted of a fluid which flowed from one body to another. Such phenomena as the temperature of a mixture of hot and cold water could be
accurately forecast from the respective quantities and temperatures
of the two portions of water. The fluid theory was in keeping with the
fluid concepts with which industry worked. That heat might be a
mode of motion was an ancient idea, arising from the mobility of
flames and the production of heat by friction. It was difficult, however, for thinkers in the early stages of the Industrial Revolution to
conceive of heat as a mode of motion, because they were pre-
dominantly concerned in processes such as distilling and evaporating
solutions of sugar or of salt, in mixing large volumes of fluids at
small differences of temperature. The fluid theory of heat seemed to
fit these phenomena more easily than the theory of heat as a mode of
motion. But difficulties arose in trying to apply it to the explanation
of the properties of heat engines.

Newcomen and Watt engines were used for more than a century
before the principles of their operation became widely understood.
The amount of heat put into the steam by the furnace and the amount
taken out by the condenser were in these engines very large, compared
with the small amount of heat converted into mechanical work. The
accurate measurement of this comparatively small quantity against
the background of two much bigger ones was technically difficult.

The low efficiency of the early steam engines, which concealed the
scientific facts of their operation, misled Sadi Carnot (1796–1831) in
his early theoretical investigation of the steam engine. He published
in 1824 a correct theory of the efficiency of heat engines performing a
cycle of operations, though at that time he accepted the incorrect
theory that heat is a fluid. Before he died, he perceived the true
nature of heat as a mode of motion, and made the first calculation of
the mechanical equivalent of heat. His figure was 370, compared with
the correct figure of 427. It is not known how he arrived at this
figure, which was found in his papers and not published until 1878.
Carnot unfortunately died prematurely of cholera, but his work on
heat caused Larmor to claim for him the position of being the most
gifted physicist of the nineteenth century.

The importance of having the support of the main stream of
endeavour in any age in obtaining the understanding and recognition

of discovery is illustrated by J. R. Mayer's (1814–1878) theoretical researches on heat. After qualifying as a medical doctor, his first professional appointment was on a Dutch ship bound for Java in 1840. He had with him the work of Lavoisier and Laplace on the role of heat in physiology, which they had published in 1780. When the ship arrived in Java some of the crew became ill. Mayer had to bleed them and was astonished to see that the blood was a much brighter red than it was in Europe.

As Lavoisier and Laplace had interpreted the production of heat in animals as the result of a slow combustion or oxidation of tissues, he inferred that the extra redness of the blood in the tropics must be because the body loses less heat owing to the higher outside temperature. Consequently less oxygen would be taken by the tissues from the red arterial blood, charged with oxygen in the lungs, so that the blood would still be comparatively red when it passed to the veins, and thus redder than it would have been in Europe. Such was the hint that was sufficient to spur on a man of genius to discover the conservation of energy.

Mayer then made a thorough study of Lavoisier's and Laplace's experiments on the heat produced by living organisms. He showed that the data of Lavoisier and Laplace indicated that if the human body is a heat engine, then the mechanical work done by the body is equal to the heat consumed, and one must be convertible into the other. He now reflected on phenomena in nature which might manifest this equivalence of mechanical action and heat. He assumed that when air is compressed, and thereby heated, all of the work done in the compression appears as heat in the air. It followed that the ratio of the specific heats of air at constant pressure and constant volume should therefore provide a measure of the mechanical equivalent of heat. He made the calculation, which Liebig published in his chemical journal in 1842.

Mayer's paper was ignored. The leading physicists objected to his assumptions, and could not take seriously the physical publications of a doctor in general practice. He published brilliant deductions of the role of the conservation of energy in nature, but in Germany his work was derided. Even Helmholtz, who had independently arrived at similar conclusions, failed to appreciate him. Poor Mayer was allowed to become demented through non-recognition, and his fame was not established until John Tyndall fought the intellectual case for him twenty years after his first publication, on material supplied by R. J. E. Clausius (1822–1878), the first discoverer of the second law of thermodynamics.

The principle that heat and mechanical power are equivalent, and can be converted into each other, was first acceptably proved by James Prescott Joule (1818–1889) of Salford near Manchester. He was born and lived in one of the centres of the new industrial age. His father owned a brewery, and in his boyhood he became familiar with the equipment of distilling and pumping machinery. It provided classical examples of the conversion of heat into work. Joule and his brother were sent to John Dalton for coaching, and so in his youth Joule became acquainted with a scientist of the highest class. Joule became interested in electric dynamos and motors, which had recently been invented. He met W. Sturgeon (1783–1850), the untutored private soldier who observed meteorological phenomena when out on army exercises and taught himself to be a natural philosopher. Sturgeon invented the electromagnet, and the commutator, which enabled a direct current to be obtained from a dynamo.

Joule was deeply impressed with these inventions, which seemed at first to promise a new form of power that would supersede the steam engine. The strength of an electromagnet depended on the number of turns in the wire coil round the soft iron core; it therefore seemed that a motor of immense power could be made merely by winding many turns in the electromagnets on dynamos. Joule's first paper, published in 1838 when he was nineteen, consisted of an account of a motor designed with many electromagnets, in an attempt to realize his idea. Later in the year he published exact measurements of the amount of power produced by a motor. He gave it in terms of foot-pounds per minute, which was the first recorded absolute measurement of work made for purposes of physical research. It represented a transfer of industrial engineering methods of thinking into science.

In order to discover the effectiveness of his various improvements in the design of electric motors, Joule embarked on a careful measurement of the amount of work that they did for the amount of electricity consumed. He discovered the law by which the strength of an electro-magnet can be calculated. He measured the amount of electricity consumed by measuring the amount of substances released by the current in electrolysis. He found that when the motor was driven by a constant battery its force decreased with increase of speed. He did

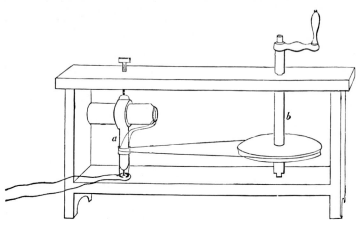

30. Joule's magneto-electric machine

not understand why, until he learned of Faraday's discovery of electromagnetic induction. It is this electrical induction which limits the speed of the motor and prevents it from becoming a machine capable of perpetual motion.

He saw that it was necessary to investigate the heat produced by the motor while it was running. As a preliminary he measured the amount of heat produced in a wire by a current of given strength, and discovered the law governing this phenomenon. In 1841, when he was twenty-three, Joule published a long account of a wonderful series of experiments in which he tracked down in detail the input and output of power and heat in a magneto-electrical machine. He suspended part of it in a tube of water, and measured the heat pro-duced by the motor by measuring the rise in temperature of the water, which rose by only one-tenth of a degree Fahrenheit.

He showed that the heat generated by electrical resistance was proportional to the resistance multiplied by the square of the intensity of the current. He adopted what he called 'a standard of resistance'. His investigation of the chemical action, current and heat produced in the electrolytic circuit, which he used for measuring his current, led him to formulate the relations between the strength of the electric current and the number of atoms involved in the chemical reactions from which it arose. The production of heat by the magneto-electrical machine led him to reflect that it would not be surprising if heat is a form of vibration, that the movement of a coil of wire before the poles of a magnet should lead to the manifestation of heat; it would merely be the transformation of one kind of motion into another.

Joule then went on to measure the power required to drive his magneto-electrical machine, by attaching weights to a string wound round its axle. Knowing the equivalent in heat of the electric current produced by the machine, and having carefully measured every way in which heat could be consumed or lost in the apparatus, he was able to calculate that the amount of heat needed to raise one pound of water one degree in temperature was equivalent to a mechanical force capable of raising a weight of 896 lb through a perpendicular height of one foot. Joule's well-known experiment, in which he rotated a paddle in water and measured the rise in temperature caused by friction, gave the more accurate result of about 782 lb. He forecast that 'we shall be able to represent the whole phenomena of chemistry by exact numerical expressions, so as to be enabled to predict the existence and properties of new compounds'. He concluded in 1844 that 'the specific heat of a body is proportional to the number of atoms in combination divided by the atomic weight'. Hence 'the zero of temperature is only 480 degrees Fahrenheit below the freezing point'. He thereby conceived and evaluated the absolute zero.

Joule described his results at various meetings, but they were received with scepticism until 1847, when he spoke at the Oxford meeting of the British Association. William Thomson (1824–1907), who had recently been appointed professor at Glasgow at the age of twenty-two, came to hear him with the intention of pointing out his mistakes. But after hearing him he presently became converted. William Thomson did for Joule what Clerk Maxwell did for Faraday.

By 1851 he had become convinced that Joule's work must be accepted, and he discovered that it could be combined with Carnot's cycle, on the basis of the conservation of energy, and the principle that heat is a mode of motion. Thus Thomson founded independently the new science of thermodynamics; he subsequently learned that Clausius had made the same discovery in Germany in the previous year.

However, German science was less ready than British to appreciate it. Clausius's work did not begin to have its full effect until the industrial and scientific development of Germany in the second half of the nineteenth century. Then this powerful force, building on his foundation, carried the German development of thermodynamics beyond the British, leading to the discovery by Max Planck (1858–1947) in 1900 that energy is discontinuous, and exists only in quanta, or portions, of finite size. The theory of this phenomenon, and its effects in nature, are expounded in the Quantum Theory.

Thomson applied the new science of thermodynamics to the elucidation of many natural phenomena. One example among many was its application in 1853 to transient electric currents, in which he explained the oscillatory character of the electric discharge from a Leyden jar. Thomson's paper was the starting-point of Clerk Maxwell's mathematical development of electromagnetic theory, and it led to the experimental discovery of radio waves, for it was presently perceived that the electric oscillations in the discharge should cause radio waves in space.

Thomson and Joule collaborated in experimental researches which led to the discovery that when gases are expanded a cooling effect may arise from the separation of the constituent molecules, if these have a slight attraction for each other. This effect was very convenient for liquefying air, and so became the foundation of the liquid air and liquid oxygen industry, and of low-temperature refrigeration.

William Thomson, later Lord Kelvin, combined an inventive engineering talent with immense scientific ability. He was born in Belfast, and matriculated at the University of Glasgow at the age of ten. Shortly after being elected to the Glasgow professorship in 1846 he showed in a short paper that electrical and magnetic forces could be represented by distortions in an elastic solid. He perceived that this contained the clue to a theory of electromagnetism but did not work it out. Clerk Maxwell picked up the clue and worked it out.

Electrical theory became of great practical importance in the design and operation of Atlantic cables. Thomson became interested, and elucidated the electrical principles which should govern the design. He showed that owing to the effects of the specific inductive capacity of the cable, it could signal quickly only if it worked with very small currents. He invented the mirror galvanometer to detect these currents, and thereby enabled the Atlantic cables to work successfully. The mirror galvanometer established a new standard of sensitivity in scientific instruments.

The work on the cable stimulated Thomson's interest in the establishment of exact standards of electrical measurement in general. He inspired the British Association to take up this question. Under its auspices much work was done, and electrical standards, such as the exact size of the ampere, volt and ohm, were established; this was essential for the development of electrical engineering and the mass production of electrical goods.

Thomson's interest in the Atlantic cables drew him into the problems of navigation. He invented improved ship's magnetic compasses. He collaborated with instrument-makers in Glasgow who manufactured his instruments. This acquainted him with the problems of the engineering development of an invention. His navigational interests stimulated him to consider the theory of the tides. He and Joule organized a collection of tidal data from different places on the British coast. His elder brother, James Thomson, who was professor of engineering at Belfast, invented a mechanical analogue computer for handling these data, which was the forerunner of the computers devised by Vannevar Bush at the Massachusetts Institute of Technology and D. R. Hartree's at Manchester University.

After the conservation of energy had been discovered it became possible to conceive clearly what happens in gases, and to work out their physical properties in detail. Foremost in this were Clausius and Clerk Maxwell, who developed the kinetic theory of gases. Maxwell showed how the molecules of a gas could be dealt with on a statistical basis, like a crowd, without having precise data on each individual molecule.

M

As the theory of energy became more refined, difficulties began to appear in it. According to the early theory, nearly all the energy emitted from a very hot body should be in the form of very short waves. Experimental investigations proved that this was not so. Ultimately Planck showed that the phenomenon could be explained if it were assumed that the mechanism of radiation in a hot body were in some way limited. A vibrating atom was apparently forced to distribute its output of waves among a comparatively limited number of wavelengths. From this it followed that energy was always emitted in packets or quanta of finite size.

After Planck proposed the Quantum Theory of Energy in 1900 scientists started to apply it to phenomena which had previously been inexplicable. Einstein showed that the new theory would explain the observed values of the specific heats of substances of low temperatures, which departed widely from the values expected on earlier theory. Then he applied it to explain the photo-electric effect, in which electrons are emitted from certain metals when they are exposed to light. The next great application of the quantum theory was by Niels Bohr (1885–1962), who in 1913 showed how the atom, as conceived by Rutherford in 1911, could behave in the manner which had been discovered from experimental physics.

The fact that energy is always manifested in packets, or quanta, led to the application of statistics to it, just as this theory had been applied to the behaviour of the molecules of gases. As the statistical theory of energy developed, it gradually became evident that the principle of probability, which underlies the theory of statistics, entered fundamentally into the properties of matter. When this was recognized the hitherto inexplicable phenomenon of the spontaneous disintegration of atoms became intelligible; it was a consequence of the operation of the principle of probability in the innermost structure of nature.

When Einstein explained the observed absolute velocity of light, which is independent of the velocity of its source, by means of his theory of relativity he pointed out that it was an immediate consequence of his theory that mass is a form of congealed energy; and that the amount of energy in absolute units in any mass is equal to that mass multiplied by the square of the velocity of light. The velocity of light is a large number, so the amount of energy congealed in mass is very large.

Thus the quantum theory of energy and relativity theory made clear the enormous store of energy in matter. This raised the question of the availability of this energy to man. Could he tap even a small part of this enormous store? As for the universe at large, the presence of prodigious amounts of energy in the matter spread in it provided an inexhaustible supply of power, by which the evolution of the universe could be explained. As a subsidiary detail in the history of the universe, the origin of the solar system and the earth, and of life on the earth, began to become intelligible.

But at the same time the scientific concepts which made the interpretation of nature more intelligible became less and less like the notions of common life. Heat, mechanical force, electricity and mass were seen as manifestations of something more abstract, that is, of energy. No image of energy in itself could be summoned up in the mind; only different manifestations of it. The notions of movement and position, drawn from experience with common objects, were found to apply less and less precisely to entities moving very fast, or to the very small.

It gradually became evident that matter and waves had a dual nature. From one aspect they appeared to consist of waves, and from another of particles. This seemed paradoxical, until it was seen that the belief that the very small, the very fast or the very big would behave like the objects of common experience was an assumption. After this had been realized it became evident that the new quantum and relativistic conceptions were just as reasonable in relation to the experiences on which they were based as are the common notions of wave and atom, space and time, to the things of ordinary experience, such as the waves on the sea, particles of dust in the air, a stone falling to the ground or a man walking from one place to another.

19. Chemistry and industry

The use of exact weighing as a means of determining questions of principle in chemistry was effectively introduced by Joseph Black and perfected by Lavoisier. Black's invention of quantitative chemical analysis by weighing the products at various stages of a chemical process was the parallel in chemistry of the balance sheet in trade. Lavoisier's systematic quantitative analysis of chemical reactions corresponds to the balance sheet in finance, which reveals the most important points in the conduct of a commercial organization. Lavoisier's personal involvement in finance and manufacture was a factor in this chemical achievement.

The social forces which caused the French Revolution gave a great stimulus to science. Before the Revolution, France had fallen into a backward condition where the difference between the aristocracy and the rest of the population was extreme. The professional classes particularly resented their situation, and put themselves, especially the lawyers, at the head of the discontented masses. When the old régime was overthrown the professional classes assumed leadership. They began to reorganize the state according to their own ideas. Professors, teachers, scientists and engineers desired that their professions and activities should have more influence, and be fostered by the new state. Plans were drawn up for the reform and promotion of science. These were to be carried out on the basis of a rational system of measurement and units; the decimal system was developed for this purpose. Education was rationalized in order to provide the necessary officials, craftsmen, engineers and scientists. The Academy of Sciences was reconstituted on a more comprehensive plan, as the Institute of France.

The scientific effects in every direction were great, but most spectacular in the mathematical sciences. This was because many of the able young men had hitherto read classics as a preparation for the legal profession. They now tended to find scope for their aptitude for abstract thought in mathematics.

Though the French Revolution was political rather than industrial, it nevertheless gave an impetus to the development of French industry. This advanced but was unable to keep up with British industry, which then had advantages in supplies of raw materials and was supported by governments that came more and more under the influence of industrialists.

One of the prime needs of the new industrialism was large and cheap supplies of soda, which is used in many cleansing and manufacturing processes. The old French Academy of Sciences offered a prize in 1775 for a method of making soda out of common salt. This was solved by N. Leblanc (1742–1806). Leblanc kept his process secret, but when France became isolated in the revolutionary wars the demand for soda became acute. He was compelled to reveal his process and was put in charge of a factory for operating it. He was not successful and in 1806 he committed suicide. Meanwhile, a British manufacturer visited Paris during the peace of 1802 and learned the process. A factory utilizing the process was successfully established near the immense salt deposits in Cheshire by J. Muspratt (1793–1886) in 1823, whose firm ultimately became part of the Imperial Chemical Industries.

As the revolutionary wars stimulated the French to work on the Leblanc process, the Franco-Prussian war caused them to search for a substitute for butter during the siege of Paris, and this led to the invention of margarine by A. Mège-Mouriès (1817–1880). He prepared a butter substitute by heating beef fat with the gastric juices from stomachs of pigs and sheep, in an alkaline solution.

The growth of the textile industry created a big demand for bleaching agents. In 1785 the French chemist C. L. Berthollet (1748–1822) suggested using chlorine for this purpose. It was used industrially at Aberdeen in 1787. James Watt recommended the father of his second wife, James MacGregor, the Glasgow dyer and bleacher, to adopt chlorine bleaching, which had been demonstrated to him in Paris by Berthollet. C. Tennant (1768–1838) of Glasgow facilitated chlorine bleaching in 1799 by absorbing the gas in slaked lime, thus inventing bleaching powder, which was much easier to handle than the gas.

The growth of printing by steam-driven machinery so increased the demand for paper that the former supply made from old rags could not meet the demand. Wood-pulp was introduced as a substitute, but this needed strong bleaching to make it sufficiently white for paper. A large new demand for chlorine bleaching thereby arose. The source for meeting it was found in the waste from the Leblanc process for making soda. In the first stage of this process common salt is heated with sulphuric acid. This causes a large quantity of hydrochloric acid to be formed. In the earlier period of the Leblanc process this was discharged into the atmosphere, or into rivers, where it caused widespread damage. In 1866 W. Weldon (1832–1885) set up an industrial process in Lancashire for obtaining the chlorine required for bleaching from the then unwanted hydrochloric acid. It consisted of heating the acid with pyrolusite, a mineral rich in manganese dioxide.

The Leblanc process for making soda was largely superseded by the Solvay ammonia-soda process. This is based on the chemical reaction, discovered by the French scientist A. J. Fresnel (1782–1827) in 1811, between common salt and ammonium bicarbonate in water. He tried to develop it industrially, but was unable to control the reaction, which is delicately balanced. The reaction was rediscovered by Ernest Solvay in 1861, and he succeeded in operating it practically. In 1872 J. Brunner and L. Mond secured a licence from Solvay to work the process. They started a factory in Cheshire and began producing soda in 1875. Their firm subsequently became a nucleus of Imperial Chemical Industries.

The early chemical processes for manufacturing chlorine have been superseded by electrolysis. This was applied after 1885, when improvements in power plants provided supplies of cheap electricity. Provision of supplies of pure chlorine led to the manufacture of hydrochloric acid, in the form of gas, by the burning of chlorine in an atmosphere of hydrogen.

The most important industrial chemical is sulphuric acid. It was known at least as early as the eighth century AD. It was distilled from iron pyrites, and the process was not improved upon until 1740, when J. Ward (1695–1761) introduced the combustion of sulphur and nitre under large glass jars containing some water. He reduced the price of the acid by a factor of about twenty. Roebuck improved Ward's process by replacing the comparatively small and fragile glass jars with large lead chambers. This put sulphuric acid production on to a completely industrial basis, and was a major factor in promoting the Industrial Revolution. The next important advance was based on the proposal by P. Phillips in 1831 to make the acid by passing a mixture of oxygen and sulphur dioxide over platinum powder. Such catalytic methods of making sulphuric acid did not become industrially successful until the latter part of the nineteenth century, owing to the poisoning of the catalysts by impurities.

The third fundamentally important industrial acid is nitric acid. It was certainly known to the Arabians in the eighth century. They made it by heating a mixture of nitre, blue vitriol and alum. The blue vitriol and alum produced sulphuric acid, which reacted with the nitre, producing nitric acid. J. R. Glauber (1604–1670) introduced in about 1648 the preparation of nitric acid by distilling nitrates with sulphuric acid. The nitrates were imported from India, where they accumulated in the huge dumps of organic refuse. This source was presently superseded by imports of nitrates from Chile. The provision of cheap electricity at the end of the nineteenth century led to the production of nitric acid by the Birkeland–Eyde process. Latterly nitric acid has been made on a large scale from the ammonia produced by catalytic methods from hydrogen and nitrogen, first worked out successfully by F. Haber.

The manufacture of chemicals used in bulk, such as the mineral acids and alkalis, is known as the heavy chemical industry. It developed first on a large scale in Britain, to supply the means for working the materials produced by the Industrial Revolution.

The critical event for nineteenth-century chemistry occurred when the youthful genius Justus von Liebig (1803–1873) was sent by his patron to study chemistry in Paris in 1822. He was born in Darmstadt, the son of a druggist. His extraordinary talent was brought to the notice of the local duke, who assisted his education. When Liebig arrived in Paris he was befriended by the famous German explorer, scientist and diplomat Alexander von Humboldt (1768–1859).

From the original stimulus of Lavoisier, the French chemists had made considerable progress in the chemistry of materials from plants and animals. This branch of chemistry is known as organic, because of the source of its materials. Lavoisier had believed that the oxygen inspired in breathing decomposed fluids passing through the lung, releasing carbon and hydrogen. These gases then combined with more inspired oxygen, and produced the carbon dioxide and water found in respired breath. Lavoisier looked for simple reactions between molecules of carbon, oxygen and hydrogen to explain what he observed in the body. But carbon and hydrogen were not found as such in the body. The researches of F. Magendie (1783–1855) presently advanced the knowledge of the chemical composition of meat, fat and bread. He determined their contents, of carbon, nitrogen, hydrogen and some other elements, fairly exactly.

After his return to Germany, Liebig pursued research in the chemistry of living organisms and their materials. By 1842 he had shown that the body does not work on simple compounds of carbon and hydrogen, nor on raw food, but on materials of an intermediate order of complexity. What, then, were they exactly? Liebig provided this department of chemistry with a programme, which has led to the recognition of proteins, carbohydrates and the modern chemical conception of the constitution of living things. Together with this exercise of the imaginative intellect, he made the further great contribution of creating new experimental technique, to which reference has already been made.

Detailed knowledge of the intricate chemistry of living things began to advance at a greatly increased rate. Liebig developed his new methods in his laboratory at the University of Giessen, where organic chemistry in the modern sense of the term was effectively born. Young men of talent from all over Europe flocked to Giessen.

As has been explained in Chapter 15, Liebig perceived that certain chemical molecules proceeded through a cycle from the inorganic world to plants and then to animals; and from the decay of plants and animals back to the earth. He pointed out that coal consisted of fossilized plants, and probably contained many substances which had not yet decomposed into simple compounds of carbon, hydrogen and oxygen. Coal should be a source of fairly complicated intermediate compounds, derived from the complex substances of living organisms but not yet broken down into the simplest substances. He thereby pointed the way to the creation of the huge organic chemistry based on the coal-tar industry.

The British industrialists were primarily interested in the heavy chemical industry in acids, alkalis and other chemicals used in bulk. These substances were comparatively simple, and elementary inorganic chemistry went a long way to meeting British needs. The organic chemistry based on materials of plant and animal origin dealt with more complex substances, which were sensitive to heat and

31. View of Liebig's laboratory at Giessen

reagents, and easily decomposed. Their general characteristics were very different from the metals, minerals, acids and alkalis whose chemical resistance and behaviour was comparatively strong or violent. The new organic chemistry became distinctly separated from the earlier inorganic chemistry, and it led to a new kind of chemical industry, manufacturing more complicated and subtle substances allied to the constituents of living organisms. This new chemical industry became known as the fine chemical industry.

Liebig found that the German agriculture and industry of his day was not sufficiently advanced to take advantage of his new chemistry. He therefore looked to England for its adoption and development.

The rulers of England in the 1840s, headed by the Prime Minister, Sir Robert Peel, whose family had been among the founders of the textile side of the Industrial Revolution, were keenly interested in learning from Liebig how science could be applied with more effect to agriculture and industry. They invited him to visit England, where he made triumphal tours, assisted in particular by his gifted Scottish student at Giessen, Lyon Playfair (1818–1898).

As a result of this activity, the Royal College of Chemistry was founded in London in 1845, with one of Liebig's most talented pupils, A. W. Hofmann (1818–1892), who had also married one of his nieces, as director. He held this position for eighteen years. During this period he had a series of outstanding pupils, which included W. H. Perkin, Henry Bessemer, Warren de la Rue, F. A. Abel, E. C. Nicholson, C. B. Mansfield, G. Merck, P. Griess, William Crookes and E. Frankland. W. H. Perkin (1838–1907) became one of his pupils at the age of fourteen. At eighteen he discovered the first synthetic dye made from coal-tar, and founded the synthetic dye-stuffs industry. Henry Bessemer (1813–1898) invented the method of making cheap steel in quantity which was fundamental for industrial progress, and particularly for the development of the United States of America.

In 1863 Hofmann decided to return to Germany. He became the inspirer of the German dyestuffs industry. He and his pupils widened and deepened Perkin's discovery, and within a score of years the German synthetic dyestuffs and fine chemical industry had out-stripped the British. As Hofmann pointed out, German national habits were particularly suited to the prosecution of organic chemistry. This was because there is scope in organic chemistry for a great deal of fruitful routine work, many slight variations of one experiment having to be performed in tracing the properties of closely related substances. Such work lends itself to organized and directed research.

In Germany the Liebig tradition produced more well-trained academic chemists. At first they found few openings in the backward German industry, and so many came to England. They obtained posts as chemists in English chemical factories, whose directors were making so much money that they did not bother about the improvement of processes. The German academic chemists, after their service in the English factories, returned to Germany and founded small German firms, making products which were an improvement on those produced by the English. By about 1880 these small firms, which from the first had been owned and directed by chemists with good academic training, had prospered greatly and began to amalgamate into the huge German chemical trusts that became so famous, and up to the end of the First World War dominated the fine chemical industry. The success of the German fine chemical industry stimulated the development of the chemical industries in Britain, the USA and the USSR.

The great increase in knowledge of organic substances inspired by Liebig created a need for new ideas to bind this knowledge in a consistent theory. The major contribution was made by F. A. Kekulé (1829–1896), who suggested in 1858 that carbon had four valencies, or links, by which other atoms could be hitched to it, and that it had also the power of combining with itself. This led to the introduction of the still-used graphic formulae, which are a strong aid to the chemist's imagination. In 1865 Kekulé suggested that the molecule of benzene, the coal-tar constituent which had provided the basis for the new synthetic dyes, consisted of six atoms forming a hexagonal ring. The idea came to him in a dream while he was dozing in front of a fire. He saw carbon atoms dancing in his imagination.

'My mental vision rendered more acute by repeated visions of the kind, could now distinguish larger structures of manifold conformation: long rows, sometimes more closely fitted together; all twining and twisting in snake-like motion. But look! What was that? One of the snakes had seized hold of its own tail, and the form whirled mockingly before my eyes. As if by a flash of lightning I awoke; and this time also I spent the rest of the night working out the consequences of the hypothesis.'

These conceptions were followed in 1874 by the idea of valencies or links pointing in different directions in space, conceived by Van't Hoff (1852–1911) and Le Bel (1857–1930). They suggested that the four links pointed towards the corners of a tetrahedron, at the centre of which was the carbon atom. This made possible the explanation of chemical properties which depended on structures as well as composition. By 1900 the organic chemists had discovered 100,000 substances, which could be visualized and chemically manipulated with the aid of these ideas.

The further developments of the twentieth century have depended on new theories of chemical combination, on the development of more powerful methods of chemical analysis and synthesis, and on the application of new physical ideas and methods to chemistry.

As mentioned in Chapter 15, chromatography is one of the outstanding recent developments in chemical analysis. Tswett, in his original work, separated the constituents of a solution of plant substances in petroleum ether by pouring them down a tube packed with calcium carbonate. A series of coloured zones appeared, each zone containing a particular constituent. A very convenient modern form of the technique, perfected by Consden, Gordon and Martin, uses blotting-paper, on which coloured spots appear in various places according to the particular substance which concentrates there. It gives in two days results which took two years by the older methods, enormously increasing the rate of progress. The rapid advances in biological chemistry in recent years have owed much to chromatography, which will deal with very small quantities of substances and greatly increases the subtlety of analysis.

Another advance, depending, like chromatography, on the phenomenon of molecular absorption is the ion-exchange analysis, introduced in 1935 by B. A. Adams and E. L. Holmes. The development of X-ray analysis has enabled the chemical structure of many important substances, such as vitamins, to be determined, and thus provided the chemist with the main clue to their synthesis. Together with these, experimental physics has provided the electron microscope, by which large molecules can actually be seen; and the mass-spectrograph, which can separate and identify extremely small quantities of substances.

Among the new techniques of chemical synthesis, the use of very high pressures has led to such discoveries as polythene. This was the result of researches pursued by A. M. J. F. Michels (born 1889) in Amsterdam, who observed that the gas ethylene tended to polymerize, that is, its molecules joined together in bigger units at very high pressures. Finally the application of thermodynamics and quantum theory has carried the understanding of chemical reactions to a yet more refined stage, opening the way to still more varied and subtle synthetic achievements.

20. Electric power

The discoveries of Volta, Ørsted, Faraday and others led to the creation of electrical engineering by inventors who were concerned with utilizing electricity for practical purposes. There was no sharp distinction between these two types of scientist; some, like the three mentioned, were almost entirely concerned with discovering the laws of nature; some, like William Thomson (Lord Kelvin), distinguished themselves both as philosophical investigators and inventive engineers; and others were primarily concerned with invention, incidentally making important contributions to science.

One of the most outstanding of this latter kind was Thomas Alva Edison (1847–1931), who probably did as much as anyone to found modern electrical engineering, through the creation of an electric-light system. The existence of such a system stimulated the development of every component in it, from the electric light to the transmission lines and the big dynamo. The demand for economy, and for large efficient machines to supply electricity in quantity, inspired the systematic application of research to every aspect of design and manufacture. Problems which had hitherto been studied in a desultory fashion, according to the particular interests of isolated investigators, were now taken up with urgency, to be solved as quickly as possible in order to satisfy public demand for electricity.

Edison's immersion in every aspect of electrical engineering development led him to several discoveries of theoretical interest. One was the Edison Effect, observed in 1883, in which the escape of electricity from hot filaments was noted. It provided the principle upon which the invention of the radio valve was to be based. He invented the aerial while attempting to produce a system of electrical communication without wires, by means of the electrical field from a highly charged body. He appears to have been the first to conceive the possibility of radio-astronomy, for after the discovery of radio waves his imagination prompted him to suggest that waves of this character might be coming into the earth from outer space. In 1890 he designed experiments for detecting such radio waves from space, which were correct in principle.

No sufficiently delicate receivers were then in existence which would have enabled his apparatus to work, and so his experiments, if attempted, could not have been successful. Nearly half a century was to pass before radio waves from space were discovered by K. Jansky (1905–1950), who, like Edison, was an American scientific engineer, concerned at the time in the development of electrical communications.

The discovery of the electric current immediately suggested the possibility of electrical communications. This was the most important early stimulus to electrical engineering. The first electric telegraph was constructed by the great mathematician and theoretical physicist C. F. Gauss (1777–1855), but it was used for the registration of magnetic observations, not for utilitarian purposes. Joseph Henry (1799–1878) in America improved the electromagnet, and devised an electric telegraph in which it was incorporated. These enabled a much bigger signal to be obtained from a small current, and thus from a long distance. He also used his electromagnets for operating levers at a distance. Henry refused to patent his inventions.

W. F. Cooke (1806–1879) and C. Wheatstone (1802–1875) constructed the first practical electric telegraph, making use of Henry's contributions, in 1837. Such a telegraph was needed for the operation of the rapidly growing railway system. They had set up a demonstration transmitter at a suburban station near London. It happened that a fleeing murderer boarded a train at this station. His pursuers telegraphed a message to London to have him detained. This was done successfully, and secured for the new telegraph, which had hitherto been regarded as a toy, serious public recognition. The American artist and painter S. F. B. Morse (1791–1872) introduced the recording electric telegraph in 1844. He invented his celebrated code for this purpose.

While the growing railway system stimulated the development of the electric telegraph in England, its expansion in America was even more compelling. Between 1850 and 1860 the mileage of American railways expanded from 7,500 to 30,000. Much of it was through unsettled country without alternative forms of communication. The virgin nature and the size of America made the electric telegraph even more important there than in England. It provided swift communication of intelligence which helped to bind the various states into a unit and had a fundamental role in converting the United States into a single nation. Before the railways and the telegraph North and South were virtually independent. But after their construction the ends of the continent were brought into closer contact, so that the conflicts inherent in their different forms of society were exacerbated.

Thomas Alva Edison was born at Milan, Ohio, a village on a canal connecting the Eastern cities of the United States with Lake Erie. His ancestors were of Dutch descent. They were independent in character; some branches of the family had sided with the Americans in the War of Independence, and others the British. Edison's branch had supported the British in that war and had had to emigrate to Nova Scotia. Edison's father then showed the family independence in supporting the rebellion of W. L. Mackenzie in Canada in 1837 against the British. He had now to flee to the United States, and it was this that had brought him to Milan.

Thomas Alva was just entering adolescence when the Civil War started. He became exposed to the tremendous pressures of that major event at the most formative period of his life. At about the age of eleven he read a simple book on physics and chemistry, or what would now be called general science, which aroused his interest, and he began to make experiments. He said that he always felt himself to be more a chemist than a physicist or engineer. He acquired two hundred bottles and various chemicals from the local drug-store, and materials for making voltaic batteries. Like other boys he made pocket-money by doing odd jobs such as delivering vegetables. At the age of twelve he aspired to sell newspapers and sweets on local railway trains. He began to work on the trains between Port Huron and Detroit in 1859, just before the outbreak of the Civil War. He bought chemicals and apparatus in order to make experiments in chemical analysis, some of which he used to make in the train van during slack periods. During the long waits at Detroit he read technical literature in the local library.

He noticed the intense interest in news generated by the Civil

War, and so started to print a small newspaper on the train – the first of its kind in the world. As the Civil War moved to its climax, public interest in war news rose to a frenzy. In 1862 the crucial battle of Shilo occurred. Edison conceived the idea of telegraphing the headline news of the result ahead of his train, to attract buyers for his newspapers. When the train arrived, there were crowds waiting for the papers with the full story; Edison sold thousands of copies. He recorded afterwards that he then 'realized that the telegraph was a great invention'.

He had already read an account of Morse's telegraph system in his science book. The telegraphists on the railways were independent and colourful personalities. There were few of them because of the war call-up, and those available could behave almost as they liked, as they could not be replaced. To their generation they were what pilots and space-men are to our own. Edison had ambitions to be a telegraphist and secured a post when he was fifteen as a substitute for a man who had enlisted in the US Telegraph Corps. He found the work hard, especially the sending of press reports, which was made more difficult by his slight deafness. While in this job he acquired much old equipment which was no longer wanted, including several ounces of platinum in disused Grove nitric acid batteries. He was still using some of this platinum in experiments in the latter part of his life. He made his first invention when he was sixteen. It consisted of a timing device which replied automatically to the hourly signals coming in to check whether night operators were awake.

The shortage of operators enabled the remainder to get jobs almost anywhere. Edison travelled over the central states between Detroit and New Orleans as a telegraphist for five years, until he was twenty-one. In 1864 he was at Indianapolis, where he invented a machine for recording press reports and then playing them back slowly. It consisted of a disc of paper with the records of the signals marked by indentations along a volute spiral. The machine contained the germ of his invention of the phonograph, or gramophone. This arose from his accidental observation that when the repeater disc was rotated the rattle of the needle against the indentations produced a musical note. Thus he was led to the idea that the human voice ought to be recordable by indentations on a suitable surface.

N

At this time Edison was uncouth, unpopular and always experimenting with equipment, trying to improve it and make the telegrapher's work, with the crude and unreliable apparatus, less difficult. He invented a functionalist form of handwriting, so that fast messages could be written down directly in script which anyone could read. After the Civil War he met an operator who had been engaged in telegraphic intelligence work, who suggested to Edison that he should invent a system of telegraphing which could not be effectively tapped. In the course of trying to solve the problem, he invented his system of quadruple telegraphy, in which four messages were sent simultaneously over one wire. It was his first important invention. Duplex telegraphy, by which two messages were sent simultaneously over the same wire in opposite directions, had already been invented. Edison devised a method of sending two messages simultaneously in the same direction. He then combined it with the duplex to obtain a quadruplex system. He introduced this system into practice in 1874. He said that the mental effort of thinking it out damaged his memory.

In 1868 he worked in Boston for the Western Union telegraphic service, and while there made his first patented invention. This was an electrical system for registering votes in the U S House of Representatives. When he demonstrated it in Washington politicians explained that the last thing they wanted was quick voting, as it made obstructive tactics more difficult. Edison henceforth decided to confine his inventive efforts to things that were wanted. One of these was a ticker for brokers' offices, which registered by telegraph the changing prices of stocks. He devised for this purpose a simplified form of printing telegraph, which had been introduced by D. E. Hughes (1831–1900) in 1855. He went to New York to sell his stock-ticker, but without success. By this time he had become entirely absorbed in invention, and deeply in debt through the expenses of experimenting. In 1869 he obtained a job with Western Union. He could not afford lodgings, and so slept in the battery room of the company, which specialized in telegraphing to speculators the variations in the price of gold. While there he studied and thoroughly mastered the company's equipment.

In the financial chaos after the Civil War, with thousands of millions of paper dollars in circulation, New York became the scene of the greatest gambling frenzy that had so far ever occurred in the world. The railway and financial magnates battled for the control and exploitation of America. Vanderbilt conceived the plan of buying all the railways serving New York, and then holding the city's trade, and most of that of the continent, to ransom. Jay Gould and Jim Fisk conceived the plan of cornering all the available gold, which had increased enormously in value owing to wartime inflation in the United States, bribing relations of the President, General Grant, in the course of their operations.

As these financial battles swayed this way and that, the prices of stocks oscillated between giddy heights and alarming depths. The telegraphic indicators of the price of gold were unable to keep up with the flood of price-changes, and the instruments jammed. A crowd of hysterical messengers arrived from Wall Street, shouting for explanations of why no prices were coming through. The president of the company rushed in, demanding that the apparatus should be put right, but the engineer-in-charge was too flustered to do it. Edison was standing near by and said that he thought he knew what should be done, and the president shouted: 'Fix it, fix it!' Edison carried out the repairs in two hours, whereupon he was appointed general manager of the system, at what to him was then the enormous salary of $300 a month.

He had difficulty in keeping the system going in the mounting frenzy of gold speculation, but he succeeded. The frenzy reached its peak on Black Friday, when some speculators actually went mad; the stock exchange had to be closed and all deals cancelled, because they had become inextricably confused. In the final hours Edison sat on top of a telegraph booth, watching the frenzied crowd, in which financiers who had lost their reason had to be physically restrained. One of the Western Union men came to him and said: 'Shake, Edison, we are O K. We haven't got a cent.' Edison said he felt very happy that they were poor, for such occasions were very enjoyable for poor men, and occurred rarely.

Edison now set up with his friend W. L. Pope a firm which they advertised as 'electrical engineers'. This appears to mark the foundation of electrical engineering as a profession. The firm produced improved tickers, and Edison invented a unison stop, by which all tickers on the system could be simultaneously brought back to the zero position, so eliminating the need to send mechanics to synchronize the instruments in different offices. To his astonishment he was offered $40,000 for this and his other ticker inventions. He used this money to start on the invention and manufacture of electrical equipment on a considerable scale. This was in 1870, when he was twenty-three. Among his first workmen were S. Bergmann and J. S. Schuckert, who later founded two of the biggest electrical trusts in Germany. Another was J. Kruesi, who became chief engineer of the General Electric works at Schenectady. Later on he employed the eminent electrical engineer, A. E. Kennelly (1861–1939), the co-discoverer of the Kennelly–Heaviside layer. Edison had the power of commanding and keeping men of outstanding ability.

He worked very intensely on the development of automatic, in place of the original hand-operated, instruments. One of the difficulties in working instruments at high speeds arose from the effects of self-induction, which caused signals to be drawn out and lose their precision. He overcame this by using induction to produce a momentary reversal of current, which cut off sharply the chemical marks of the recorder. He received an award for this at the US Centennial Exposition of 1876, on the adjudication of Sir William Thomson, who described it as 'a very important step in land telegraphy'.

Alexander Graham Bell patented his invention of the telephone in 1876. He utilized an iron diaphragm, which was made to vibrate by the human voice in the field of a magnet and thus produce a current which corresponded in strength to the variations in the voice. The original telephone could be used only over short distances, as the current produced by the vibrations of the diaphragm was very weak. Edison almost immediately made two inventions which turned the telephone into a widely practical instrument. One of these was the carbon transmitter, by which the signal current could be amplified, so that long-distance telephony became possible. The device consisted of a carbon button which pressed on the diaphragm. When the latter vibrated under the voice the resistance between it and the button varied. Consequently, if the button were placed in the primary circuit of an induction coil and the distant receiver in the secondary circuit of the coil a high-voltage current which varied according to the voice could carry the signal to the distant receiver. The resistance of long-distance wires could thereby be overcome.

The second invention was a non-magnetic relay, which made the Western Union Company, with which he was working, independent of electromagnetic relay patents. It was based on his discovery that moistened chalk became slippery when a current passed through. Consequently a lever held at rest by friction on the chalk could be released when a current went through it. Edison then applied this invention to make his company independent of the Bell receiver, which consisted of a diaphragm caused to vibrate by the signal current, the reverse of the Bell transmitter. He arranged a metal rod to rest on a cylinder of moistened chalk. An end of the rod rested against a mica disc. When the chalk cylinder was rotated and a signal current went through the moistened chalk the rod began to slip and caused the disc to vibrate. The disc then caused the air to vibrate and reproduce the sounds of the voice which had originally produced the signal current. This remarkable instrument was a loudspeaker, the energy being provided by the rotation of the cylinder. When George Bernard Shaw was a young man he worked as a salesman for the Edison Company in London. He had to demonstrate this chalk-cylinder loudspeaker, and he found it a fearsome piece of apparatus. It was successful, however, in extricating the Edison Company from dependence on that of Bell.

In 1876 Edison set up a laboratory of a new type at Menlo Park, a house in the country about twenty-five miles from New York. Its aim was to pursue invention to order, for industrial and commercial purposes. He undertook to try to invent any kind of device that was desired or seemed to meet a demand. The invention of the carbon telephone transmitter was his first major work at Menlo Park. Then in 1877 he invented the phonograph, or gramophone, based on the adaptation of the mechanism of his automatic recording telegraph to another purpose.

After the triumph of the gramophone he looked for a new major field. It was suggested to him that he should invent a practical small electric light, suitable for homes and offices. The electric arc light, invented by Humphry Davy, had been made commercially successful, but it operated only in powerful, flickering units, with a palling colour. It was useful for lighting public places, but not for the domestic and office market. When he returned to the electric lamp after the gramophone Edison made a study of the economics of illumination, in order to discover what characteristics an electric lamp must have to compete successfully with gas. He saw that it should be a high-voltage, low-current lamp, for this would minimize the cost of the copper conductors in the transmission system. At the same time the voltage should not be too high, or it would be dangerous. The radiating properties of various shapes of filament were calculated, including the effect of close coiling, by which one part of the filament obstructed the radiation of another (the principle underlying the 'coiled coil' filament, introduced more than half a century later).

It was this quantitative and developmental research on lamp design which enabled Edison to overtake others in the same field. J. W. Swan (1828–1917) had made a carbon electric lamp at Newcastle in 1860, but it had only a short life. Progress did not become possible until H. J. P. Sprengel (1834–1906) introduced his mercury vacuum pump in 1865. Swan resumed research on carbon electric lamps in 1877. He patented the method of heating the carbon filament while the bulb was being exhausted, in order to drive out the occluded gases. Edison completed his lamp in 1879. Swan's filament heating patent would have prevented the introduction of Edison's lamp into England, and so Edison and Swan combined to make the Ediswan lamp.

In his intense research on carbon filaments, Edison drew upon his experience in working with carbon in his invention of the carbon transmitter. He tried six thousand varieties of vegetable fibres as the source of carbon filament, and found bamboo to be the best. In his efforts to improve the carbon lamp he investigated the blackening of the inside of bulbs. He noticed that blackening was less where the plane of the filament loop intersected the inside surface of the bulb. He concluded that atoms of carbon were being shot off the filament and some were being stopped by other parts of the filament, which was obstructing their flight. He made a bulb with a metal plate placed between the legs of the filament and attached to a wire sealed through the base of the lamp. If the positive leg of the filament were connected to the plate a small current passed through the wire,

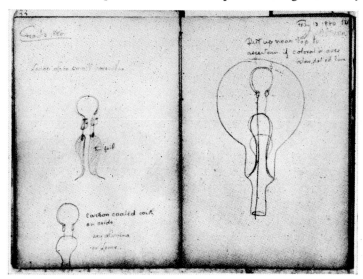

32. Edison's notes on the 'Edison Effect'

whereas if the plate were connected to the negative leg no current passed. Thus the arrangement would allow a negative but not a positive current to pass. The phenomenon became known as the Edison Effect.

At this time, in 1883, J. A. Fleming (1849–1945) was on Edison's staff in London. He started systematic research on the Effect, and based his invention of the radio valve on it. He perceived that it could be used for obtaining direct currents from the oscillatory currents set up in radio receivers. In 1907 Lee de Forest (1873–1961) added his invention of the grid, which made the valve amplify as well as rectify.

The invention of a practical and economic electric lamp opened up the prospect of large-scale demand for electric power. Edison now directed his attention to the invention of large electricity generators and transmission systems, with their multitude of incidental devices. He designed large dynamos with low internal resistance, introducing mica laminated armatures, and he invented insulating tape. John Hopkinson (1849–1898), the professor of engineering at Cambridge in England, developed the theory of dynamo design, which enabled defects in Edison's original designs to be eliminated. Edison and Hopkinson independently invented the three-wire system, which reduces the amount of copper cable required in transmission.

Edison used direct current in his original electric-light system because he was familiar with it. The problems of using alternating current, which has many advantages in principle, were more difficult and had not yet been worked out. The success with direct current increased confidence in attacking the problems of alternating-current supply. The development of central electricity generating stations producing electric power led to the creation of the heavy electrical industry, and the production of electricity as a commodity. In the previous telegraphic development the cost of electricity had been a minor factor; in power supply it became a major one.

Edison supervised the new factories for manufacturing lamps, dynamos, cables and fittings for his system. He devised electricity meters based on electrolysis, and through these sold units of electric power to consumers. Edison held a thousand patents, most of which were for electrical devices. His company became the embryo of the General Electric Company, and his research laboratory the original of the famous research laboratories of the great electrical corporations.

21. Scientific method in industry

Scientific investigation accustomed men to being systematic and exact, for experience showed that this made discovery easier. After the notion of a scientific method had become clear and success had given men confidence in it they began to apply it to the improvement of industry, and to apply industrial methods to science.

These developments were at first most spectacular in the invention of new processes based on deeper physical and chemical knowledge, which occurred at the beginning of the Industrial Revolution. They were followed by the application of science not only to the invention of new processes but also to the more efficient management of those already known, including the operations of manufacture as well as the production of a wider range of materials and larger sources of power.

By the latter part of the eighteenth century and the beginning of the nineteenth there were already a number of notable examples. The great French physicist C. A. Coulomb (1736–1806) applied scientific methods in the organization of armaments manufacture. Matthew Boulton junior and James Watt junior, the sons of the great engineers, made considerable advances in the methodical layout of factory machinery, and in time-and-motion study to discover the most economical way of performing a craftsman's operations. Marc Isambard Brunel (1769–1849), the father of Isambard Kingdom Brunel, organized the manufacture of pulley-blocks for the British Navy on mass-production principles, involving the analysis of the operations of manufacture and their division into more than forty individual processes, each of which could be performed by a special machine designed for the purpose.

The theoretical implications of these developments were grasped by Charles Babbage (1791–1871). About the year 1813, when he was a student at Cambridge, he perceived that it ought to be possible to

mechanize mathematical calculations. This led him to invent a series
of calculating machines, ending with one that was theoretically
capable of solving any mathematical problem to any reasonable
degree of approximation. He discovered the fundamental principles
of the modern computer, and even invented the computer termi-
nology which is used today. As James Watt applied science to the
machine, Babbage applied the machine to science. He saw that
research in science would ultimately become the solution of problems
by machinery. He forecast in 1838 that the application of the computer to atomic theory would reduce chemistry to a branch of mathematics. He said that, given the properties of atoms, it would ultimately be possible to calculate from their properties all the kinds of
substances, and their properties, which might exist.

Babbage was the son of a banker. He was brought up in the
managerial and accounting atmosphere, and absorbed its characteristic attitudes. He showed mathematical ability and was sent to
Cambridge. Being rich he had no need to adapt himself to the traditional Cambridge intellectual interests which, in science, still
reflected those of the pre-industrial age represented by Newtonian
astronomy.

Babbage and the other most talented men of his generation,
J. F. W. Herschel (1792–1876) and G. Peacock (1791–1858), looked
to French science for more modern ideas. French mathematics
especially had been stimulated by the French Revolution. Among the
many scientific reforms that were carried out was the establishment
of the decimal system of measurement. This involved the recomputing of the mathematical tables used to facilitate calculations of many
kinds. A committee of leading mathematicians under de Prony
(1755–1839) was appointed to recommend how the enormous work
of recomputing the tables should be done. While he was on a vaca-
tion, Prony happened to read Adam Smith's *Wealth of Nations* and
his descriptions of the subdivision of labour which was being introduced in industry. It occurred to him that the recomputing of the
tables on a decimal basis might be facilitated by a similar subdivision
of labour. The chief mathematicians indicated the principles, based
on the method of differences, on which the computations might be
subdivided into a series of elementary arithmetical additions and

subtractions, that could be carried out by unskilled persons who knew no mathematics beyond simple arithmetic. About a dozen competent mathematicians were engaged to work out the details of the sub-divisions, and about eighty unskilled persons to do the additions and subtractions. The huge tables were recomputed very quickly by the team.

Babbage was fascinated by this work. While reflecting on it, he perceived that the ultimate steps in the calculations, already reduced to additions and subtractions, might be made entirely mechanical and be performed by a machine. He set about inventing such a machine,

33. Part of Babbage's computer

which he called a *difference engine*. After this success he began to think about the kind of operations carried out in mathematical analysis, and how they might be done by machinery. He perceived that a machine which consisted essentially of two parts, like the warp and weft in weaving, might be made to do it. He even saw that the punched cards used in the Jacquard loom for weaving designs could be employed to control the two operations. He gradually thought out the complete principles of the modern computer, and gave the names, such as *store, mill, memory*, to its parts and properties. He set about designing a computer which, in principle, was as mathematically powerful as any that has since been made.

His mind went at once to the limit of what could be required by any reasonable scientific problem. He tried to make such a machine, but it was beyond the engineering resources of his day, which were restricted to rods and wheels. It did not become practicable until the almost infinitely versatile electron could be harnessed to do the work, instead of clumsy rods and wheels. This was first accomplished about a hundred years later, under the pressure of the Second World War, to facilitate calculations for military weapons. With the aid of radio valves and, later, transistors, the electron was made to carry out the operations which Babbage had envisaged with rods and wheels. The electronic machine was much smaller and enormously faster.

Babbage perceived that calculation would be applied in an increasingly comprehensive way to the operation of machinery, and to the whole industrial and commercial process. In his *Economy of Machinery*, published in 1828, he showed how the principles of calculation could be used to discover the most efficient way of carrying out an industrial process. This was an example of what today is called *operational research*. He foresaw the principles of scientific industry of the future. Babbage promises to be to the age of automation what Newton was to the age of navigation.

22. The application of mathematics to biology

The application of mathematics to the analysis of industrial operations has opened the prospect of a new automated industry. Its application to the analysis of the properties of living things has given a new insight into their structure and mechanism. The progress of science is always stimulated when the facts and phenomena under investigation are successfully brought under mathematical description, because the processes of mathematics may be used to deduce what the descriptive formulae imply, leading to new knowledge and suggesting fresh problems for investigation, which often could not have been perceived by direct inspection. When the facts under investigation are small in number and capable of precise definition mathematics may be extremely successful in describing them, and enabling deductions to be made as to how they will work. This is seen in Galileo's original mathematical analysis of motion, leading to the successful calculation of the range of projectiles.

The great success of mathematics in dealing with mechanics and physics led to the belief that measurement and successful application of mathematics was the mark of genuine, or at least of mature, science. This attitude was expressed by Lord Kelvin, who did not feel satisfied with scientific knowledge until it had been stated in measurements and subjected to calculation.

The prestige of science which had been given a mathematical expression caused investigators to try to apply mathematics to biology, in the hope of deriving a complete account of the biological world, parallel to Newton's physical world. These attempts had little success, and their authors claimed too much for them, so that biologists came to distrust the application of mathematics to biology.

One of the fundamental difficulties was to discover biological phenomena which were sufficiently precise to be susceptible to the kind of mathematical treatment that was applied so successfully to cannonballs and iron rods. Another of the difficulties was that biological phenomena are so multitudinous and complex.

This led L. A. J. Quetelet (1796–1874) to apply the methods of mathematical statistics to the behaviour of living organisms, both human and non-human. By these methods it was possible to discover regularities among the enormously variable data of biological phenomena. His work inspired Francis Galton (1822–1911), a cousin of Charles Darwin, to apply statistical methods to a variety of biological phenomena. He obtained mathematical formulations of many regularities appearing in the descendants of man, horses, dogs and cattle, and tried to explain them in terms of his cousin Charles's theory of heredity. Darwin had accumulated much information on heredity, both by experiment and reading, but his theory of heredity was defective. He believed that the main, though not the exclusive, feature in the emergence of new varieties and species was the accumulation of small differences, or continuous variation. As this was not correct, Galton was unable to discover the mechanism of heredity, but he did succeed in describing mathematically the variation and likenesses appearing in relatives, in the case of those characters which varied continuously. His work in this direction was carried forward with the aid of more advanced mathematics by K. Pearson (1857–1936) and W. F. R. Weldon (1860–1906).

At the very time that Galton was pursuing his application of statistics to biology, Gregor Mendel (1822–1886), an Augustinian monk at Brno, was establishing by theory and experiment the data on which a mathematical theory of heredity could be adequately based. During the previous hundred years a great deal of data on the breeding of flowering plants had been collected by gardeners, who supplied new plants for the ornamental gardens of landowners and improved vegetables for market gardeners. The Augustinian monks were keenly interested in plants, for increasing both the beauty and profitability of their considerable estates. Mendel was himself the son of a small farmer; it was therefore not difficult for him to become interested in gardening. He devoted himself to the grafting of plants, plant-breeding, the cultivation of bees and meteorology.

As a student of ornamental plants, he was aware of the rough arithmetical rules of plant breeders for estimating the frequency with which particular types of plant would appear in the offspring of hybrids – $\frac{1}{2}, \frac{1}{4}, \frac{1}{8}$ etc. He became impressed with the constancy of the inheritance of characteristics in peas. It struck him that experimental breeding with the pea might lead to the discovery of the mechanism of inheritance. In 1854 he started a series of breeding experiments, which continued for about a decade and of which he published accounts in 1866. He seems to have formulated a correct theory of inheritance on the basis of his general knowledge of pea-breeding before he started on these experiments, conceiving that heredity is transmitted by units or factors which are independent, and can pass through many generations unchanged. Particulate or atomistic theory lends itself to mathematical description, and Mendel had a sufficient grasp of mathematics to appreciate this.

He saw that if the heredity of the pea were of this nature, then, if two varieties of peas were crossed which differed only in one character, this character would appear among the first generation of offspring in numbers governed by a simple formula. For instance, the character might appear in one-quarter of the offspring, but not in the remaining three-quarters; that is, in the ratio 1 to 3. If the plants to be crossed differed in two characters, then these might appear in the offspring in the ratios 9 : 3 : 3 : 1; with increasingly complex but calculable distributions for larger numbers of different characters.

Mendel conducted many breeding experiments with peas, in which these mathematical distributions were confirmed with remarkable exactitude. He understood the implications of his work, and recognized its bearing on Darwin's theory of evolution. His account of his results, published in the local natural history journal in Brno in 1866, was virtually ignored. Brno was then in the Austrian Empire, away from the contemporary centre of the scientific world, in England, Germany and France. The ideas of Darwin were more in tune with flourishing industrial Western Europe. Scientists did not look to Brno for discoveries. During the next thirty-four years many of Mendel's results were gradually rediscovered, independently, but published simultaneously in 1900 by H. de Vries (1848–1935), C. E. Correns (1864–1933) and E. Tschermak (1871–1962).

When the atomistic nature of heredity had been recognized biologists looked for its material basis. The coloured particles called chromosomes, found in the nuclei of the cells of living organisms, seemed the sort of structures which might have such a role.

T. H. Morgan (1866–1945), H. J. Muller (1890–1967) and C. B. Bridges (1889–1938) began experiments with the banana fruitfly, *Drosophila melanogaster*, in 1911, which breeds quickly and has easily distinguishable chromosomes. They accumulated a great deal of evidence that the Mendelian factors resided in the chromosomes, and the development of their work enabled the location of many of these factors in the chromosomes to be determined. The Mendelian factors, now called genes, were identified in increasing detail and numbers, until they began to share similarities with the small differences postulated by Charles Darwin.

In 1918 R. A. Fisher (1890–1962) showed that the mathematical results obtained by Pearson and Weldon were a necessary consequence of the Mendelian theory. The conceptions of Galton and Mendel were gradually brought together through the development of sufficiently refined experiments and statistical analysis. The Galtonian approach revealed the average behaviour of all the genes involved in the inheritance of a character, while the Mendelian gave information about individual genes; the Galtonian was more comprehensive but less specific than the Mendelian.

When the Mendelian theory was in its earlier stage of development it seemed difficult to reconcile with the Darwinian theory of natural selection, for the material factors of inheritance appeared to be permanent. Mendel was aware that his theory presented problems to the contemporary Darwinian theory. Later research showed that the material Mendelian factors were subject to occasional mutations, or sports. In 1930 R. A. Fisher demonstrated that organisms whose genes were subject to such mutations would, under the effects of natural selection, follow a biological evolution of the kind observed in nature. Thus Mendelism and Darwinism were reconciled.

Much has since been learned of the character of the gene. It consists primarily of deoxyribonucleic acid, DNA, with a helical structure discovered by J. D. Watson (born 1928) and F. H. C. Crick (born 1916) in 1953, which has the property of splitting along its length, the two halves having the power of assimilating from their environment fresh molecules, so that each half again becomes a

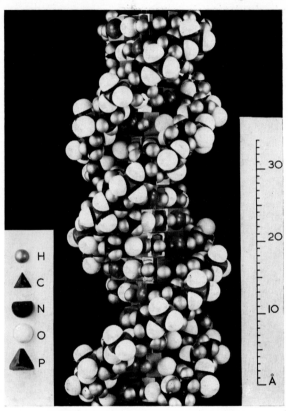

34. The DNA molecule

complete helix. On this splitting mechanism, the transmission of heritable characters is based.

The mechanism of inheritance turns out to be similar to the programme which is fed to a computer in order to enable it to solve a problem. The genetic mechanism in the living cell programmes the chemical material in its environment, so that this is built up into an

O

organism with specific characteristics. Thus, while the computer with its 'memory' and its anticipatory powers becomes more and more like a living organism, the inner mechanisms of living organisms appear to be more and more like computers. Living organisms and machines seem to be developing along converging lines, and in the future one principle may be found underlying both.

23. The atom

The progress of chemistry, physics and electricity has enabled individual atoms to be identified. Research on their properties has led to the discovery of how they can be artificially disintegrated and how atomic energy can be released. The uncontrolled release of atomic energy has produced the atomic bomb, while its controlled release has provided the principle on which nuclear power stations work. The fusion of atoms offers unbounded supplies of energy. The process already occurs in the stars, and it may well be that the solution of the problem of fusion will be found as much through astronomy as through physics. In the past astronomy has helped agriculture and navigation; in the future it may prove of prime importance to industry. Already the exploration of space has given a great stimulus to research and engineering development. It promises to inspire future science, as the desire to explore the surface of the earth inspired science in the past.

Atomic theory has a long history. The early Greek philosophers invoked the idea of the atom to explain the variety and stability of nature. If there were not some limitation on the degree to which matter could be subdivided it was hard to see why nature should not be just an ever-changing fluid, in which nothing had any degree of permanence. Unless there were a limit, it was difficult to understand how stable things, such as stones, could exist.

As knowledge of the facts about material things and their changes increased rapidly during the Renaissance, natural philosophers were increasingly fascinated by the idea of atoms. They speculated on how physical and chemical properties and changes might be explained on the supposition that they were due to interactions between atoms. Bacon suggested that heat consisted of vibrating motions of the constituent particles of substances. Newton supposed that light consisted of subtle particles. Chemists ascribed the combustion of substances to the exchange of fiery particles. Bodies expanded when heated, because the motions of their particles became larger and occupied more space.

However, the attempt to place this atomic theory of phenomena on a mathematical basis proved very difficult. The first mathematical deduction of Boyle's Law from the concept of gases as particles in motion did not occur until 1738, when Daniel Bernoulli (1700–1782) solved this problem; and also deduced that if the pressure of a mass of gas is kept constant its volume will increase with increase in temperature. Bernoulli's deduction of the effect of temperature on gas at constant pressure was overlooked, and rediscovered experimentally by J. A. C. Charles (1746–1823), who published it in 1802. Atomic theory was largely dropped during the eighteenth century because of the difficulty of applying it with precision to the explanation of phenomena. By the beginning of the nineteenth century enough new quantitative information had been gained and put in order, especially by Lavoisier, for it to become possible to fit together atomic theory and chemical and physical facts in a significant way.

John Dalton was the chief author of this reconciliation. He was a Quaker who originally conducted a school for children at Kendal in Cumberland. He read mathematics and science, and made experiments by himself. He first became interested in atomic theory through reading Newton's *Principia*. Living near the Lake District, where atmospheric effects and scenery are striking, he was attracted to the study of meteorology. He started regular observations which he kept up for several decades, and began experiments to gain insight into atmospheric phenomena. Through his study of the effects of heat on air, he rediscovered, in 1801, the law of Daniel Bernoulli and Charles, of the effect of heat on the volume of gases at constant pressure. About the same time his consideration of the effects of water-vapour in the air led him to discover that, in a mixture of gases, each gas exerts its own pressure independently.

His chemical analyses of the air showed that it consisted of a fairly uniform mixture of nitrogen, oxygen, carbon dioxide and water-vapour. He saw that the atomic theory about which he had read in Newton might offer an explanation of his law of partial pressures, and of the uniformity of the atmosphere. The particles of the different gases in the mixtures being uncombined, they necessarily exerted their effects independently. The atmosphere, in spite of the differences in density of its constituents, was uniform because the motions of the individual particles of the gases, moving among each other, kept the gases perpetually intermingled.

Dalton's researches in meteorology and physics had thus led him to concrete evidence for the atomic theory of matter. He now proceeded to investigate how it might be applied to the explanation of the great advances that had recently been made, especially by Lavoisier and the French chemists, in determining the exact proportions in which various substances combined. This had led to the recognition of at least thirty 'elements', out of which other substances were made by chemical combination. Dalton assumed that an 'element' was made of atoms, all of which were similar and could not be divided by any known process. The atoms of each element had a weight and properties specific to that substance alone. Chemical compounds were formed by the combination of atoms of different elements in simple numerical proportions.

He explained the atomic constitutions of whole ranges of chemical substances on these lines, and invented a symbolism to represent them: the parent of the systems of chemical and atomic formulae which are still used. Dalton expounded his theory systematically in his *New System of Chemical Philosophy*, published in 1808.

Nevertheless, nearly fifty years passed before his new atomic theory of chemistry exerted its full influence. This was because he had not been able to recognize that the smallest particle of water must consist of *two* atoms of hydrogen combined with one of oxygen. A. Avogadro (1776–1856) showed in 1811 how this difficulty could be overcome, if it were assumed that equal volumes of all gases under the same conditions contain equal numbers of particles. He called such particles *molecules*. Avogadro's hypothesis was rediscovered nearly half a century later by Cannizaro, in 1854.

In the second half of the nineteenth century chemists found the concept of the chemical atom of immense help in their researches into the constitution and structure of substances, especially in the chemistry of the compounds of carbon. By the end of that century they had worked out the constitution and structure of thousands of

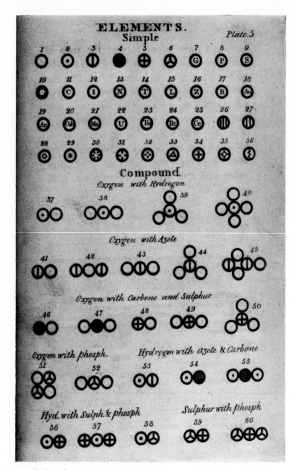

35. Dalton's chemical symbols

substances, and synthesized many of importance. This great success gave chemists the strongest belief in the existence of chemical atoms. Their confidence prompted them to believe also in the immutability of the atom. The smallest amounts of matter that could be measured by the technical means available in the nineteenth century were comparatively so large that they contained millions of millions of atoms. Thus only the properties of large numbers of atoms were directly known. The properties of individual atoms could only be inferred as averages of those measured in large numbers. In spite of this, scientists in general were convinced that individual atoms were absolutely identical and immutable.

Yet in the middle of the nineteenth century the recognition of periodic properties in the chemical elements had already provided evidence that the different kinds of chemical atoms were closely related, and were therefore probably composite structures made out of the same primary substance. The leading part in this work was taken by the Russian chemist D. I. Mendeleev (1834–1907), who in 1869 began to work out a classification of the properties of the then known elements, based on the principle that they are in periodic dependence on their atomic weights. He showed that the relations between the atoms so revealed were very extensive and complex. In particular he found that there were three gaps in his table, corresponding to atoms which were not yet known. He forecast what their properties would be. These three missing elements, gallium, scandium and germanium, were subsequently discovered in 1874, 1879 and 1885 respectively. Their properties were found to be very close to those Mendeleev had forecast. On the basis of Mendeleev's table, a number of scientists of whom Lyon Playfair was one of the foremost, forecast that chemical atoms would be transmuted.

General scientific opinion remained unshaken in the belief in the immutability of atoms until the discovery of the electron in 1897. The proof that electric particles existed with a mass less than one-thousandth of that of a hydrogen atom, and that particles of the same kind could be obtained from all the different chemical elements, made it seem possible that the different chemical atoms were merely different combinations of electrons. As these were negatively charged, they were presumably held together by some kind of positive force or particle. During the first years of the twentieth century efforts were made to locate the positions of electrons in atoms. J. J. Thomson saw that the way in which X-rays were scattered by matter should throw light on this problem. His pupil C. G. Barkla (1877–1944) used this technique to demonstrate that there were layers or shells of electrons in the atoms, and J. J. Thomson calculated the number of electrons within the atom. He showed approximately that there was a relation between this number and the chemical properties of the atom.

Meanwhile the discovery of radioactivity by Becquerel, and of radium in 1898 by Pierre Curie (1859–1906) and his wife Marie Curie (1867–1934), opened yet another revolutionary line of advance. Radium provided very much stronger sources of the new phenomenon, which greatly facilitated research. The Curies measured the amount of heat being produced by radium, and showed that it continued without any decrease in weight that they could detect. It was evident that energy was being released by radium on a comparatively enormous scale. The Curies concluded that it must be atomic energy. They found that the heat was produced by complex radiations emitted by the radium.

The precise nature of these radiations was determined by Rutherford. He showed they were of three kinds, consisting of α-particles that became atoms of helium; β-particles, which were electrons; and γ-rays, which were electromagnetic radiation. It was evident that radium atoms were disintegrating, and that the helium atoms were some of the fragments. With the collaboration of F. Soddy (1877–1956), Rutherford demonstrated that radioactivity was due to the spontaneous disintegration of atoms.

Rutherford now decided to use the radioactive radiations to explore the structure of the atom and the mode of its disintegration. His pupils H. Geiger (1882–1945) and E. Marsden (born 1889) had observed that when the α-particles, or helium radiation, from radium was directed at thin metal foil most went through almost undeflected, but a few bounced almost straight back instead. This showed that the atoms which formed the metal foil must be mainly empty structures, through which α-particles could usually shoot without deflection. But the occasional violent recoil showed that there must be a comparatively small but heavy nucleus, from which the α-particle recoiled on the rare occasions when it collided with this.

Rutherford announced the nuclear theory of the atom in 1911, thus founding nuclear physics. In 1913 his pupil Niels Bohr showed how Rutherford's nuclear atom combined with a quantum mechanism would explain a multitude of facts revealed by the science of spectroscopy. The Rutherford–Bohr atom provided a key to numerous refinements of the Periodic Classification of the Elements, which had gradually been accumulated through nearly half a century of researches by chemists.

J. J. Thomson had found evidence that there was a relation between the number of electrons in the atom and its chemical properties. H. G. J. Moseley (1887–1915) clarified this decisively in 1913. He applied the X-ray reflection method of crystal analysis, invented by W. L. Bragg in the previous year, to the precise measurement of the very short wave-radiations emitted by atoms, and was able to show that they depended on the atomic number, which was identical with the charge on the nucleus of the Rutherford–Bohr atom. Thus the chemical properties of the atom depended on its number or nuclear charge, and not on its mass. There might be atoms with different masses but the same chemical properties. Ordinary elements might be mixtures of several kinds of atoms, which would explain why the average weights of their atoms were not exact multiples of a fundamental unit.

In the meantime the investigators of radioactivity had discovered that some of the atoms produced as a result of disintegration, although they differed radioactively, were chemically indistinguishable. In 1910 F. Soddy named such substances *isotopes*, because they occupied the same place in the Chemical Periodic Classification, though they were physically different. W. Crookes had foreshadowed the notion of isotopes in 1886. The common elements were mixtures of isotopes of several different masses. F. W. Aston (1877–1945) developed the mass-spectrograph, by which the various isotopes in any chemical element could be separated, and so decisively confirmed this deduction.

In 1919 Rutherford succeeded in causing the transmutation of an atom, by bombarding nitrogen with fast α-particles. In the following year, 1920, he reviewed the subject of the *Nuclear Constitution of Atoms*, drawing forth the implications of the discoveries of the previous quarter of a century's research. He forecast the existence of the neutron and heavy hydrogen, and of hydrogen and helium atoms of mass three.

The neutron was discovered twelve years later, in 1932, by J. Chadwick (born 1891). F. Joliot (1900–1958) and his wife Irène Curie

(1897–1956) succeeded in 1934 in making radioactive substances artificially, which led the way to an immense extension in the number and strength of available radioactive materials.

The progress of nuclear research prompted Rutherford and others to encourage attempts to imitate radioactive nature by inventing machines which could accelerate atomic particles to such high speeds

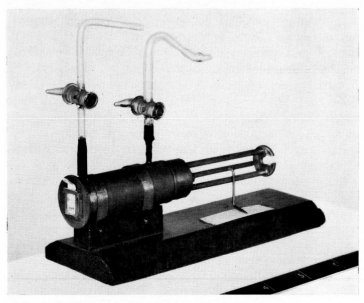

36. Rutherford's apparatus for disintegrating the atom

that they could disintegrate atoms. The first success was gained by J. D. Cockcroft (1897–1967) and E. T. Walton (born 1903) in 1932, shortly after the discovery of the neutron. Attempts were immediately made to extend the disintegration of atoms by means of neutrons and accelerated particles. In 1934 E. Fermi (1901–1954) showed that neutrons moving slowly could produce transmutations in atoms. His results were so numerous and complex, especially in the effects of neutrons on heavy elements such as uranium, that their full meaning required a good deal of analysis. Finally, in 1938, O. Hahn (1879–1968) and F. Strassmann (born 1902) showed that the uranium atom was split into two approximately equal parts by the neutron, with the release of a great deal of atomic energy. This process was elucidated by O. R. Frisch (born 1904), who named it *fission*.

Early in 1939 F. Joliot and his colleagues showed that when a uranium atom is split by a neutron more than two neutrons are released in the fission, besides the other fragments of the uranium atom. This raised the possibility of a chain reaction, by which the fission of one uranium atom led to the fission of two more, and so on. Such a chain reaction was first achieved by Fermi in December 1942 in Chicago. He succeeded in starting one in pieces of uranium surrounded by blocks of carbon, in a device called for this reason a *pile*. Fermi's pile was the first nuclear reactor. It operated with slow neutrons, under control, and was the forerunner of the nuclear power plant.

The first atomic bomb, exploded in 1945, consisted of uranium in which an uncontrolled chain reaction was promoted by fast neutrons, thus producing a sudden release of atomic energy, with enormous explosive effect. Fission bombs cannot be made more powerful than a certain size, because beyond this size the explosion blows away the uranium faster than the chain reaction can develop in it. They have been succeeded by fusion bombs, in one kind of which hydrogen combines with itself to form helium. There is theoretically no limit to the size of fusion bombs, which might be made so big that they endangered the survival of life on earth.

The energy of the sun arises from a fusion process in its interior, which results in hydrogen being converted into helium, with release of atomic energy. As the sun has been stable for hundreds of millions of years, its fusion process is automatically controlled by nature. Man aims at imitating the natural controlled fusion processes in the laboratory, so that this can be made the source of virtually unlimited supplies of power.

The first nuclear station to supply power regularly was opened in the USSR in 1954. The Calder Hall nuclear power station, opened in England in 1956, was built for the dual purpose of making radioactive material for atomic weapons and contributing a substantial amount of electricity to the English public supply system. Its originally designed output was 92,000 kW, but it was found that this could be increased considerably. Calder Hall uses natural uranium as fuel, enclosed in graphite and cooled by carbon dioxide gas. Nuclear plants of one million kW output entirely for public supply have since been built and brought into use.

The fission process offers a wide variety of different types of design, of which perhaps between ten and twenty offer practical possibilities. It may be a long time before the most convenient and economic have been determined, so that design becomes standardized.

Intense efforts are being made to discover how the fusion process can be controlled and utilized. One line of research is to contain hydrogen within a magnetic field and heat it by electromagnetic means so that fission begins. This requires a temperature of about forty million degrees. So far temperatures of the order of only about one million degrees have been obtained, and so the practical utilization of the fission process for supplying power seems still a considerable way off. As the fusion process supplies the energy of the stars, it is possible that clues to the solution of the fusion problem will come from the development of astronomy.

24. The little and the big

The first possibility of recognizing the effects of single atoms arose from the study of substances, such as zinc sulphide, which fluoresce when subjected to radiations of various kinds. When examined under a microscope the fluorescence of the zinc sulphide which is being irradiated is seen to be due to brilliant little green flashes of light. It gradually became clear that these flashes came from atoms in the zinc sulphide, which were being struck by atoms or quantum packets of wave-radiation in the irradiating beam.

Rutherford employed the zinc sulphide screen to detect the atomic fragments knocked out of the nitrogen atoms in his first demonstration of the artificial disintegration of atoms in 1919. The kind of flash seen on the screen helped to identify the nature of the atomic fragment. It was possible to see the effects of a single atom, in spite of its extremely small size, because of its great speed and very high energy.

Owing to this, the swiftly moving particle could make its presence felt in other ways. Air through which it passed was electrified, and consequently became electrically conducting. H. Geiger (1882–1945) made use of this effect by connecting a closed chamber containing air to a counter. Each time a particle passed through the chamber a momentary electric current passing through the electrified air operated the counter. In this way the number of particles passing through the chamber could be counted automatically. The Geiger counter has been, and still is, of great practical use for counting energetic atomic particles.

The most remarkable of the instruments used for detecting atomic particles is the cloud-chamber invented by C. T. R. Wilson (1869–1959) in 1911. This makes the tracks of particles visible. Moist air in a chamber with a glass window is suddenly expanded. If a fast atomic particle happens to be passing through, it electrifies, or ionizes, the gas along its track; if the conditions of humidity and pressure are appropriate water-vapour will begin to condense on the electrified particles, thus marking the track with a row of cloud droplets.

In 1925 P. M. S. Blackett (born 1897) secured for the first time a photograph of an atom being disintegrated by an impinging particle. He did this with a cloud-chamber. More recently the bubble-chamber, the spark detector and other devices have been added to the physicist's equipment for studying individual atoms and sub-atomic particles.

37. Wilson's cloud chamber

M. von Laue (1879–1960), W. Friedrich and P. Knipping demonstrated in 1912 that the atomic structure of crystals could cause the diffraction of a beam of X-rays passing through them. His discovery was extended by W. H. Bragg (1862–1942) and W. L. Bragg (born 1890). Later in 1912 the Braggs invented and developed the analysis of the structure of crystals by the reflection of X-rays from the orderly arrays of their constituent atoms.

The theory that matter possesses properties both of particles and of waves was proposed by L. de Broglie (born 1892) in 1924, and proved experimentally by C. J. Davisson (1881–1958) and L. H.

Germer (born 1896) in 1928, and G. P. Thomson (born 1892) in 1929. It followed from this theory that electrons should exhibit wave as well as particle properties. They therefore should have some properties analogous to light-waves, and be able to perform actions analogous to those of light-waves, such as those performed by light-waves in a microscope. As electron-waves were in general much shorter than light-waves, it should be possible to use them to reveal objects much too small to be visible in the most powerful optical microscope.

38. The giant accelerator at Geneva

In 1926 E. Ruska (born 1906) invented the electron microscope, and in 1931 built one, to utilize the wave properties of electrons for magnifying very small objects. It has since been developed so far that it can reveal the features of very small objects, such as the viruses which cause many diseases, and even chemical molecules of a large kind.

With all these increasing means for investigating the very small, there has been a parallel development in the investigation of the very big.

This has included the invention of machines for producing particles with very high energies. The most important contribution was made by E. O. Lawrence (1901–1958), by his effective introduction of the cyclotron in 1933. This consists essentially of the giving of repeated electrical impulses to particles restrained to revolving in a circle by a magnetic field. Developments of this type of machine, such as that at CERN, the European Centre for Nuclear Research at Geneva opened in 1960, produce particles travelling so fast that they possess energies of 30,000 million electron-volts. Plans for machines producing 300,000 million electron-volt particles are being discussed. The effects can be enormously enhanced by arranging that such particles, travelling in opposite directions, should collide.

While these advances were made in atomic and sub-atomic particles with high investigating energies, comparable advances were made in probing the depth of space. The largest optical telescope yet made for this purpose is the 200-inch telescope at Mount Palomar in California (Fig. 20).

Modern cosmology was founded by William Herschel (1738–1822), who suggested that certain nebulae were distant 'island universes', consisting of disc-shaped assemblages of stars like our own galaxy, or the Milky Way.

The distance of the nearest of the 'island universes', the nebula in Andromeda, was determined from the detection in it, by E. P. Hubble (1889–1953) in 1925, of certain stars of a type discovered by Henrietta S. Leavitt (1868–1921) in 1905, which vary precisely in brightness and whose luminosity is exactly related to their periodicity. The distance of the Andromeda nebula could be inferred at once from the apparent magnitude of the stars of this kind located in it. This nearest galaxy proved to be of the order of a million light-years away. Assuming that galaxies are of approximately the same size, their distance could be deduced by comparing their brightness with that of the Andromeda nebula. Thus it was ascertained that some of the faintest nebulae must be more than a thousand million light-years distant.

Hubble announced in 1929 his discovery of the red-shift in the spectra of distant nebulae. This was due to their swift movement away from the earth, causing an apparent increase in the wavelength of their emitted light, in the same manner that the pitch of an engine whistle becomes lower after the engine has passed the observer. Hubble's studies of the red-shifts led him to conclude that the speed of recession of the nebulae was in direct proportion to their distance. Thus the measure of the red-shift became a measure of the distance

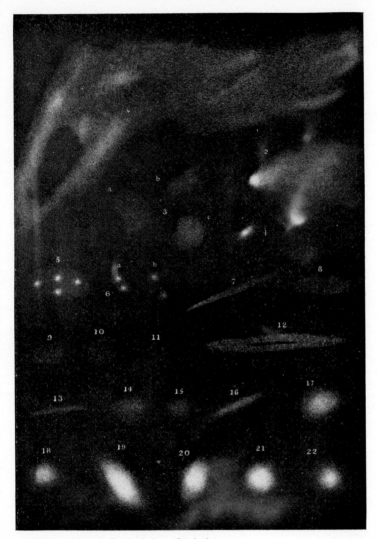

39. William Herschel's depiction of nebulae

P

of the nebulae. The red-shift for certain nebulae was so large that it indicated that the whole nebula must be receding with a speed which was a considerable fraction, a fifth or more, of the speed of light.

The next major penetration into the universe came from the development of radio-astronomy. This arose from the discovery by K. G. Jansky (1905–1950) in 1932 that radio waves from outer space were coming into the earth. Jansky was a radio engineer, engaged in research on atmospherics in order to discover how to mitigate their effects on radio transmission. Further advances came unintentionally from the development of radar for detecting enemy aircraft during the Second World War. Sensitive radar receivers were made to pick up the weak reflections of radio waves from aircraft. It happened that these worked on about the same wavelength as most of the cosmic radio waves. In 1942 some British military radar equipment was found to be subject to jamming. It was at first thought that this must be due to the enemy. Investigation by J. S. Hey showed, however, that the waves which were causing the jamming came from the sun. In 1946 I. S. Shklovsky (born 1916) suggested that the waves were produced by the motion of electrified particles in the sun's magnetic field.

Among the physicists engaged in radar during the war were A. C. B. Lovell (born 1913) and M. Ryle (born 1918). Lovell returned to academic work in 1946 with the intention of trying to detect by means of radar the clouds of electrification, or ionization, in the upper atmosphere caused by cosmic rays. He constructed a stationary parabolic reflector 218 feet in diameter. This proved so successful that he designed the great manoeuvrable radio-telescope, with a bowl 250 feet in diameter, which was opened at Jodrell Bank near Manchester in 1957. Ryle at Cambridge designed a radio-telescope on interferometer principles, analogous to Michelson's optical interferometer telescope. In 1952 he built a telescope on these lines, with four aerials placed at the corners of a rectangle 1,900 feet long and 168 feet wide. Ryle quickly increased the number of recognized radio sources in the sky from about 100 to 2,000.

Radio waves can bring information from greater depths of space than can optical waves, because the recession of their source reduces their intensity less than that of optical waves. The light from a very distant galaxy is lengthened so much by the red-shift that it cannot affect a photographic plate at all. Radio waves from the same object, however, though lengthened correspondingly, would still be observable.

Ryle added another ingenious device to his telescopes: 'aperture synthesis'. This was the use of a computer to build up a complete radio picture of a place in the sky from the collection of partial observations provided by the interferometer telescopes. In 1965 he brought into commission a still bigger telescope. This has three aerials, each with sixty-foot bowl reflectors. They are arranged so that two of the aerials can be held at 2,500 feet apart, while the third can be moved into successive positions along a railway 2,500 feet long. With this instrument he has detected a radio galaxy which appears to be 8,000 light-years distant.

These giant radio telescopes have given indications to the astronomers what to look for with their giant optical telescopes, which are operating at the limit of their power and are handicapped in observing very distant objects. As a result of radio observations it was discovered that certain distant galaxies were emitting radio waves with great intensity. Optical study of some of these revealed that they appeared to be in turmoil, as if they consisted of two galaxies in process of colliding, and through this generating radio waves very powerfully.

Then it was discovered, in 1963, that certain of the distant sources emitting radio waves intensely were comparatively very small in total volume. They were more like single super-stars than diffuse collections of stars. They produce energy with enormous intensity, possibly by processes new to physics. These quasi-stars were named *quasars*. In 1965 A. R. Sandage (born 1926) discovered, with the Mount Palomar 200-inch optical telescope, objects like quasars, which do not, however, emit radio waves intensely. The extreme brightness of these 'quiet quasars' enables them to be seen optically at very great distances, of the order of thousands of millions of light-years.

One way of explaining the recession of the nebulae is to suppose that the present universe is the result of a total explosion occurring perhaps 12,000 million years ago. The universe was then very much smaller, and all the matter in it was packed into this small volume.

The galaxies are fragments from the primeval explosion. Those which started off with the biggest velocity have got farthest, and the light and radiation from them has taken longest to reach the earth. It follows that these very distant nebulae, as they appear to us, show us what galaxies looked like perhaps 10,000 million years ago. As there is evidence that the primeval explosion occurred about 12,000 million years ago, these very distant objects, such as the 'quiet quasars', show what the galaxies and the universe were like only 2,000 million years after they were born; that is, at an early stage in their life.

Such were some of the notions being inspired in 1965 through measurement of big distances and big energies, with the aid of ever-bigger scientific instruments. The construction of such instruments depends in turn on the economic and technical resources of those who make them. Thus the grandest discoveries about the properties and structure of the universe depend upon, and cannot be separated from, the intellectual, economic and technical resources of man, and in the last analysis depend upon the totality of his development as an individual and social being.

25. Space

Increasing mastery of the land, water and air of the earth has led to the possibility of leaving the earth, and exploring and utilizing outer space. This has posed a series of very difficult problems of how to move through space and how to live in it. Their solution will require intense scientific efforts, which may lead to discoveries in every branch of science as striking as any that have been made on earth.

Space offers the possibility of converting the earth into man's home, and transferring all dangerous activities, such as the production of atomic energy, to the moon and other celestial bodies. It is conceivable that the power so produced could be transmitted across space to the earth for the use of mankind, by a device which concentrates energy in a narrow beam of radiation, somewhat after the manner of a laser, which can send extremely narrow and intense beams of light.

The idea of flying through space is one of the oldest of human dreams. It did not appear such a difficult thing in ancient times, when the universe seemed to be only a small extension of the earth, and the atmosphere was supposed to extend to its not very distant limits. Aristarchus (310–250 BC) discovered how to calculate the size of the earth and the distances of the moon and sun, but his discoveries were not generally accepted until modern times.

Galileo's telescopic observations of the moon showed that it had features not unlike some of those found on the earth. Kepler tried to deduce the origin of these features. He suggested that the lunar craters had been excavated by beings on the moon to protect them from the heat of the sun's rays. The observation of concrete features on the moon strengthened the belief that it was inhabited, and increased the desire to go there. Kepler's researches had, however, convinced him that space must be empty, so that wings would be of no use as they depended on the resistance of the air. It must also be very cold. Man could therefore neither fly with wings nor breathe, and would be frozen on his journey.

As with so many parts of physical science, Isaac Newton made an enormous advance in the understanding of the problem of penetrating space. His clarification of the principle of action and reaction being equal and opposite provided a precise understanding of how a body could be propelled by a rocket. He himself suggested the use of rockets for propelling vehicles, and it was evident from the principle of action and reaction that a rocket could be propelled in empty space by means of the jet issuing from it.

His theory of gravitation made it possible to calculate the speed that a projectile must have to escape from the earth and evolve in an orbit round it, thus becoming an artificial satellite. He described how this might be done in his *System of the World*, a draft of a section which he wrote for the *Principia* but did not include because he considered it too popular. It was published in 1728, the year after his death. He supposed that a cannon ball was projected in a horizontal direction from the top of a high mountain, where the air was rare and its resistance to a moving object could be neglected (Fig. 21). If the velocity were sufficient the cannon ball would not fall on the earth at all but go round it, and return to the mountain-top whence it had started. 'If we now imagine bodies to be projected in the directions of lines parallel to the horizon from greater heights, as of 5, 10, 100, 1,000 or more miles, or rather as many semi-diameters of the earth', then these bodies 'will describe arcs either concentric with the earth, or variously eccentric, and go on revolving through the heavens in those trajectories, just as the planets do. . . .'

This was accompanied by a diagram of the paths of such artificial satellites. Newton settled the mechanical principles of space-travel; the accomplishment of it was thenceforth a technological and biological problem.

The Chinese invented the rocket at least seven hundred years ago. It was probably the development of the fire-arrow, shot into wooden fortifications to set them on fire. Rockets were used against the British in India in the eighteenth century. This led Congreve (1772–1828) to improve the rocket as a weapon by systematic research. It seemed that the rocket would supersede the gun, but this did not happen soon because the technical advances engendered by the Industrial Revolution enabled the problems of gun-making to be solved before those of rocket-making. Research on rockets was therefore left to men off the conventional line of scientific and technical development.

The most distinguished of these was K. E. Tsiolkovsky (1857–1935), a Russian schoolmaster working at Kaluga, far from the centre of European scientific development in the nineteenth century. He attacked the problems of space travel mathematically and scientifically. In 1895 he published a paper in which he described how the astronaut would have to travel in a sealed cabin, with purification apparatus and oxygen supplies. He analysed the characteristics of a rocket capable of leaving the earth. In 1903 he suggested that liquid fuels, such as paraffin, would provide twice as much energy as solid fuels. The Rumanian mathematician H. Oberth (born 1894) carried forward the theory of space rockets in a book published in 1923.

The drive to solve the great technical problem of making big liquid-fuel rockets arose out of the results of the First World War. The Versailles Treaty laid down that the German Army was not to be allowed to possess big guns. The German General Staff therefore sought for an alternative, and it decided in 1929 to look into the possibilities of rockets, which were not covered by the Treaty.

In 1932 they engaged Werner von Braun (born 1912), a young student of astronomy with interests in space travel, and the engineer W. Riedel, to work on liquid-fuel rockets. By 1934 rockets driven by alcohol and liquid oxygen were successfully fired to a height of more than one mile over the North Sea. A big research station was then constructed at Peenemünde on the Baltic coast, starting in 1936. The first successful big rocket was fired on October 3rd, 1942, on the day after Fermi had started the first nuclear pile in Chicago. It travelled 125 miles. Bombardment of Britain with V2 rockets based on this model began in 1944.

The development of big rockets on the lines started in Germany has led to the subsequent launching of earth satellites, at first carrying scientific equipment and experimental animals. Among other results, these have revealed the existence of zones of electrified particles in regions around the earth, called the *van Allen belts*. The activity of

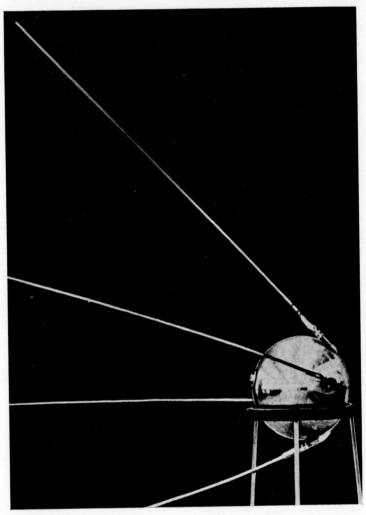

40. Sputnik I

electrified particles in the whole volume of the solar system is of a far greater complexity than had formerly been suspected, and it is possible that an understanding of it will throw light on electrical and other physical conditions on the earth.

The far side of the moon has been photographed from a rocket successfully guided round it, and the pictures radioed back to the earth. Close photographs of the planet Mars have been radioed from probes which have carried instruments near it. A number of astronauts have already travelled in artificial satellites around the earth and have been brought back, as well as sent up, safely.

The primary motive for the development of rockets has been, and still is, military. They are capable of delivering hydrogen bombs at any place on the earth. They send artificial satellites carrying instruments which will photograph the country over which they are passing, and transmit the photographs to their base. They establish satellites which remain over a definite place on the earth and act as relays for telecommunications, offering an immense extension of such facilities. In the leading industrial countries space research and development absorbs a substantial part of the national scientific, technological and industrial effort.

What the exploration of the earth was to Columbus and the men of the Renaissance, the exploration of space is to the men of today. The modern problem is, however, far more challenging, for it would seem that it will be very difficult to settle and make use of the moon, the planets and other cosmical objects, for the benefit of mankind. However, the more difficult the problem to be solved, the greater will be the influence on mankind of its ultimate solution. As Tsiolkovsky has said: 'The Earth is the cradle of the mind, but one cannot live for ever in a cradle.'

Index